Brain–Computer Interfaces 1

Series Editor
Maureen Clerc

Brain–Computer Interfaces 1

Foundations and Methods

Edited by

Maureen Clerc
Laurent Bougrain
Fabien Lotte

ISTE

WILEY

First published 2016 in Great Britain and the United States by ISTE Ltd and John Wiley & Sons, Inc.

ISTE Ltd
27-37 St George's Road
London SW19 4EU
UK

www.iste.co.uk

John Wiley & Sons, Inc.
111 River Street
Hoboken, NJ 07030
USA

www.wiley.com

Library of Congress Control Number: 2016942416

British Library Cataloguing-in-Publication Data
A CIP record for this book is available from the British Library
ISBN 978-1-84821-826-0

Contents

Foreword

A Brain–Computer Interface (BCI) records brain signals, translates them into commands that operate a variety of devices, and provides feedback to the user about how intentions are transformed into actions. These three essential components, forming a closed-loop system, define the core components of a BCI. Their natural target population has traditionally been people with motor disabilities that have lost control of their body but have preserved cognitive functions, and BCIs have been intended to act as alternative assistive devices for them. However, in recent years the scope of a BCI has widened to include restoration or rehabilitation of motor and even cognitive functions for patients after some kind of central nervous system injury, brain state monitoring for healthy subjects, and new tools for studying human brain functions.

An anecdotal, even fringe, field of research at the confines with science fiction when it appeared, BCIs have grown over the last 40 years from early prototypes in a handful of locations to more than 3,000 research labs and nearly 150 companies working in BCI-related areas nowadays. The complexity of today's BCI systems, which are moving beyond constrained laboratory conditions, calls for truly multidisciplinary efforts spanning clinical research to computer science and human–computer interfaces, from neuroscience to biomedical and neuroengineering, from rehabilitation to robotics and virtual reality, and from human psychology to material and electrical engineering.

This wide range of fields that contribute to BCI makes it difficult, if not impossible, to have a unified view covering all the facets of this fascinating scientific and translational enterprise. Thus, a certain bias is always present and openly acknowledged in our research. This book is no exception. It is edited by signal processing and machine learning specialists. Yet, aiming to become a reference for the French speaking research community, it gathers a collective body of expertise from all the fields involved in BCI research and practice. We consider this a challenge that the editors have successfully tackled, as the book covers state-of-the-art research and results in a way that all other communities can relate to. Furthermore, the curious layperson – I hope you are if you want to live long with a healthy brain! – can also profit from a significant number of chapters that do not require any specific background.

The book is organized into seven parts, distributed in two volumes. In the first volume (*Foundations and Methods*), readers walk along the path covering the main principles of BCI, with all its subtle meanders which they may decide to jump over or to explore in more details. This is a volume that we may well need to read in several iterations as we go into detail into the field and its different components. Part 1 provides all the necessary background in anatomy and physiology of the brain and nervous system to understand BCI from a neuroscience perspective. Part 2 covers the signal processing and machine learning sides of BCI, while Part 3 deals with human learning, and the interplay between the human and the machine.

Brain–Computer Interfaces by Clerc, Bougrain and Lotte is the first BCI book for and by the French-speaking community. Here, it is also translated in English as it has important lessons for all BCI researchers and practitioners worldwide. I am certain that this book will appeal to each of them as it has done to me. Enjoy it.

José DEL R. MILLÁN
Geneva
Switzerland
May 2016

Introduction

A Brain–Computer interface (BCI) is a system that translates a user's brain activity into messages or commands for an interactive application. BCIs represent a relatively recent technology that is experiencing a rapid growth. The objective of this introductory chapter is to briefly present an overview of the history of BCIs, the technology behind them, the terms and classifications used to describe them and their possible applications. The book's content is presented, and a reading guide is provided so that you, the reader, can easily find and use whatever you are searching for in this book.

I.1. History

The idea of being able to control a device through mere thought is not new. In the scientific world, this idea was proposed by Jacques Vidal in 1973 in an article entitled "Toward Direct Brain–Computer Communications" [VID 73]. In this article, the Belgian scientist, who had studied in Paris and taught at the University of California, Los Angeles, describes the hardware architecture and the processing he sought to implement in order to produce a BCI through electroencephalographic signals. In 1971, Eberhard Fetz had already shown that it was possible to teach a monkey to voluntarily control motor cortex brain activity by providing visual information according to discharge rate [FET 71]. These two references show that since that time, BCIs could be implemented in the form of invasive or non-invasive brain activity measurements, that is, measurements of brain activity at the neural or scalp

Introduction written by Maureen CLERC, Laurent BOUGRAIN and Fabien LOTTE.

levels. For a more comprehensive history of BCIs, the reader may refer to the following articles: [LEB 06, VAA 09].

Although BCIs have been present in the field of research for over 40 years, they have only recently come to the media's attention, often described in catchy headlines such as "writing through thought is possible" or "a man controls a robot arm by thinking". Beyond announcements motivated by journalists' love for novelty or by scientists and developers' hopes of attracting the attention of the public and of potential funding sources, what are the real possibilities for BCIs within and outside research labs?

This book seeks to pinpoint these technologies somewhere between reality and fiction, and between super-human fantasies and real scientific challenges. It also describes the scientific tools that make it possible to infer certain aspects of a person's mental state by surveying brain activity in real time, such as a person's interest in a given element of his or her environment or the will to make a certain gesture. This book also explores patients' expectations and feedback, the actual number of people using BCIs and details the material and software elements involved in the process.

I.2. Introduction to BCIs

Designing a BCI is a complex and difficult task that requires knowledge of several disciplines such as computer science, electrical engineering, signal processing, neuroscience and psychology. BCIs, whose architecture is summarized in Figure 1.1, are closed loop systems usually composed of six main stages: brain activity recording, preprocessing, feature extraction, classification, translation into a command and feedback:

– *Brain activity recording* makes it possible to acquire raw signals that reflect the user's brain activity [WOL 06]. Different kinds of measuring devices can be used, but the most common one is electroencephalography (EEG) as shown in Figure I.1;

– *Preprocessing* consists of cleaning up and removing noise from measured signals in order to extract the relevant information they contain [BAS 07];

– *Feature extraction* consists of describing signals in terms of a small number of relevant variables called "features" [BAS 07]; for example, an EEG signal's strength on some sensors and on certain frequencies may count as a feature;

– *Classification* associates a class to a set of features drawn from the signals within a certain time window [LOT 07]. This class corresponds to a type of identified brain activity pattern (for example the imagined movement of the left or right hand). A classification algorithm is known as a "classifier";

– *Translation into a command* associates a command with a given brain activity pattern identified in the user's brain signals. For example, when imagined movement of the left hand is identified, it can be translated into the command: "move the cursor on the screen toward the left". This command can then be used to control a given application, such as a text editor or a robot [KÜB 06];

– *Feedback* is then provided to the user in order to inform him or her about the brain activity pattern that was recognized. The objective is to help the user learn to modulate brain activity and thereby improve his or her control of the BCI. Indeed, controlling a BCI is a skill that must be learned [NEU 10].

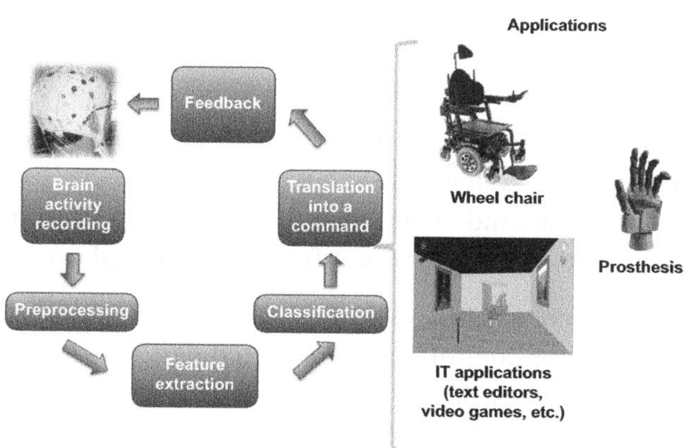

Figure I.1. *Architecture of a BCI working in real time, with some examples of applications*

Two stages are usually necessary in order to use a BCI: (1) an offline calibration stage, during which the system's settings are determined, and (2) an online operational stage, during which the system recognizes the user's brain activity patterns and translates them into application commands. The BCI research community is currently searching for solutions to help avoid the costly offline calibration stage (see, for example, [KIN 14, LOT 15]).

I.2.1. *Classification of BCIs*

BCIs can often be classified into different categories according to their properties. In particular, they can be classified as active, reactive or passive; as synchronous or asynchronous; as dependent or independent; and as invasive, non-invasive or hybrid. We will review the definition of those categories, which can be combined when describing a BCI (for example a BCI can be active, asynchronous and non-invasive at the same time):

– *Active/reactive/passive* [ZAN 11]: an active BCI is a BCI whose user is actively employed by carrying out voluntary mental tasks. For example, a BCI that uses imagined hand movement as mental commands is an active BCI. A reactive BCI is a BCI that employs the user's brain reactions to given stimuli. BCIs based on evoked potentials are considered reactive BCIs. Finally, a BCI that is not used to voluntarily control an application through mental commands, but that instead passively analyzes the user's mental state in real time, is considered a passive BCI. An application monitoring a user's mental load in real time to adapt a given interface is a passive BCI;

– *Synchronous/asynchronous* [MAS 06]: user–system interaction phases may be determined by the system. In such a case, the user can only control a BCI at specific times. That kind of system is considered a synchronous BCI. If interaction is allowed at any time, the interface is considered asynchronous;

– *Dependent/independent* [ALL 08]: a BCI is considered independent if it does not depend on motor control. It is considered dependent in the opposite case. For example, if the user has to move his or her eyes in order to observe stimuli in a reactive BCI, then BCI is dependent (it depends on the user's ocular montricity). If the user can control a BCI without any movement at all, even ocular, the BCI is independent;

– *Invasive/non-invasive*: as specified above, invasive interfaces use data measured from within the body (most commonly from the cortex), whereas non-invasive interfaces employ surface data, that is, data gathered on or around the head;

– *Hybrid* [PFU 10]: different neurophysiological markers may be used to pilot a BCI. When markers of varied natures are combined in the same BCI, it is considered hybrid. For example, a BCI that uses both imagined hand movement and brain responses to stimuli is considered hybrid. A system that combines BCI commands and non-cerebral commands (e.g. muscular signals) or more traditional interaction mechanisms (for example a mouse) is also considered hybrid. In sum, a hybrid BCI is a BCI that combines brain signals with other signals (that may or may not emanate from the brain).

I.2.2. *BCI applications*

Throughout the last decade, BCIs have proven to be extremely promising, especially for handicapped people (in particular for quadriplegic people suffering from locked-in syndrome), since several international scientific results have shown that it is possible to produce written text or to control prosthetics and wheelchairs with brain activity. More recently, BCIs have also proven to be interesting for people in good health, with, for example, applications in video games, and more generally for interaction with any automated system (robotics, home automation, etc.). Finally, researchers have shown that it is also possible to use BCIs passively in order to measure a user's mental state (for example stress, concentration or tiredness) in real time and regulate or adapt their environment in response to that state.

I.2.3. *Other BCI systems*

Let us now examine some systems that are generally related to BCIs. Neuroprostheses are systems that link an artificial device to the nervous system. Upper limb neuroprostheses analyze electric neuromuscular signals to identify movements that the robotic limb will carry out. Neuroprostheses are not BCIs if they do not employ brain activity, but rather, the peripheral nervous system activity. Exoskeletons also make it possible to bring life to a limb by equipping it with mechanical reinforcement, but to date they are very seldom activated by brain activity[1]. Cochlear implants and artificial retinas can be compared to neuroprostheses since they connect a device that replaces

1 However, the MindWalker project has started research in that direction; see https://mindwalker-project.eu/.

a defective organ with the central nervous system. However, these kinds of implants differ from BCIs in their directionality, since they do not measure neural activity, but rather stimulate it artificially.

I.2.4. *Terminology*

Several other terms are employed to refer to BCIs. In this regard, the term "brain–machine interface" refers to the same idea, although the term is more often used when the brain measurements are invasive. Although more rarely, the term "direct neural interface" is also sometimes used to designate BCIs. In this book, the term "brain–computer interface" will be employed because it underscores the idea that the processing chain is not fixed; this is to say that the system may adapt to evolutions in brain signals and the user's preferences through learning. The acronym BCI will also largely be used throughout the book, since it is the most commonly employed.

I.3. Book presentation

This book seeks to give an account of the current state of advances in BCIs by describing in detail the most common methods for designing and using them. Each chapter is written by specialists in the field and is presented in the most accessible way possible in order to address as large an audience as possible. This book, Volume 1 (Foundations and Methods), is followed by a second book, Volume 2 (Technology and Applications).

I.3.1. *Foundations and methods*

This first volume introduces the basic notions necessary to understand how a BCI works.

The brain stands at the core of a BCI. It is an organ whose functioning still remains largely beyond our understanding, although its basic principles are known. The first part of the book, entitled "Anatomy and Physiology", explains the anatomical and physiological foundations of BCIs, as well as the pathologies to which they can be applied. This part also explores devices that make it possible to measure brain activity. Finally, it studies the neurophysiological markers used in active or reactive BCIs, and in passive interfaces.

The second part, which is entitled "Signal Processing and Learning", focuses on brain activity analysis. This preparatory phase that precedes the implementation of a BCI consists of a processing chain. Preprocessing makes it possible to increase the percentage of useful signals. In turn, it becomes necessary to represent those signals in a simplified manner in terms of characteristics that are potentially useful to the BCI. According to the type of BCI, relevant characteristics will vary greatly, and two chapters will study those issues for EEG recordings, as well as for intracerebral recordings. The last crucial stage is that of machine learning, which makes it possible to define appropriate classifiers adjusted to and optimized for each user. Learning proceeds in two stages: the calibration stage generally takes place offline and operates on data gathered when the user repeatedly performs mental tasks that are relevant to the BCI, following instructions provided to him or her. Those recorded brain signals will serve as examples in order to find the best calibration settings for that particular user. Next, the online, closed loop, usage stage applies the classifier to new data.

The third part, entitled "Human Learning and Human–Machine Interaction", analyzes the BCI use phase. Machine learning must adapt to changes that can take place over a long period of time. To that end, BCIs use adaptive learning methods. Using a BCI is not a self-evident task, and we will examine the human learning that is necessary in order to attain the skills necessary to do so. Concepts of Human–Machine interaction must also be taken into account in order to best use the commands emitted by a BCI and to ensure an optimal user experience, that is, a usable, effective and efficient interaction. Finally, we will explore the concept of neurofeedback, or the perceptive feedback provided to users about their brain activity, and we will also study the relation between this approach and BCIs.

I.3.2. *Reading guide*

This book is intended for anyone seeking to understand BCIs, their origins, how they work, how they are used and the challenges they face. It may prove useful for people approaching the field in order to carry out research (researchers, engineers, PhD students, postdoctoral fellows) but also for present and future users (patients, medical practitioners, video game developers and artists), as well as for decision makers (investors, insurance experts and legal experts).

In order to facilitate the reading of this multidisciplinary book, we have provided an icon signaling the scope of each chapter's content. Chapters that are essential for understanding how BCIs work are denoted with ♥ . Those chapters compose a common core of indispensable knowledge, which can be complemented by more specialized notions in:

– neuroscience 🧠

– math and computer science ⚡

– clinical fields ⚕

– technological fields ⚙

– fields concerning societal issues ⚖

We suggest the following reading combinations according to readers' profile or to their field of specialization:

– general public: ♥

– patients: ♥ + ⚕+ ⚖

– medical/clinical practitioners: ♥ + ⚕+ 🧠

– neuropsychologists, cognitive neuroscientists: ♥ + 🧠+ ⚙

– mathematicians, computer scientists: ♥ + ⚡

– electrical engineers, mechatronic engineers: ♥ + ⚙+ ⚡

– investors, insurance experts and legal practitioners: ♥ + ⚖

I.4. Acknowledgments

This book is the collective work of a very large number of co-workers from very different disciplines, which would not have been possible without their contributions. We would like, therefore, to thank all the authors, and to all the colleagues and friends who have helped us in writing this book.

We are indebted to Flora Lotte for creating the cover illustration.

I.5. Bibliography

[ALL 08] ALLISON B., MCFARLAND D., SCHALK G. *et al.*, "Towards an independent Brain–Computer Interface using steady state visual evoked potentials", *Clinical Neurophysiology*, vol. 119, no. 2, pp. 399–408, 2008.

[BAS 07] BASHASHATI A., FATOURECHI M., WARD R.K. *et al.*, "A survey of signal processing algorithms in Brain–Computer Interfaces based on electrical brain signals", *Journal of Neural Engineering*, vol. 4, no. 2, pp. R35–57, 2007.

[FET 71] FETZ E.E., FINOCCHIO D.V., "Operant conditioning of specific patterns of neural and muscular activity", *Science*, vol. 174, 1971.

[KÜB 06] KÜBLER A., MUSHAHWAR V., HOCHBERG L. *et al.*, "BCI meeting 2005-workshop on clinical issues and applications", *IEEE Transactions on Neural Systems and Rehabilitation Engineering*, vol. 14, no. 2, pp. 131–134, 2006.

[KIN 14] KINDERMANS P.-J., TANGERMANN M., MÜLLER K.-R. *et al.*, "Integrating dynamic stopping, transfer learning and language models in an adaptive zero-training ERP speller", *Journal of Neural Engineering*, vol. 11, no. 3, 2014.

[LEB 06] LEBEDEV M., NICOLELIS M., "Brain-machine interfaces: past, present and future", *Trends in Neurosciences*, vol. 29, no. 9, pp. 536–546, 2006.

[LOT 07] LOTTE F., CONGEDO M., LÉCUYER A. *et al.*, "A Review of classification algorithms for EEG-based Brain–Computer Interfaces", *Journal of Neural Engineering*, vol. 4, pp. R1–R13, 2007.

[LOT 15] LOTTE F., "Signal processing approaches to minimize or suppress calibration time in oscillatory activity-based Brain–Computer Interfaces", *Proceedings of the IEEE*, vol. 103, no. 6, pp. 871–890, 2015.

[MAS 06] MASON S., KRONEGG J., HUGGINS J. *et al.*, "Evaluating the performance of self-paced BCI technology", Report, Neil Squire Society, 2006.

[NEU 10] NEUPER C., PFURTSCHELLER G., "Neurofeedback training for BCI control" in GRAIMANN B., PFURTSCHELLER G., ALLISON B. (eds), *Brain-Computer Interfaces*, Springer, 2010.

[PFU 10] PFURTSCHELLER G., ALLISON B.Z., BAUERNFEIND G. *et al.*, "The hybrid BCI", *Frontiers in Neuroscience*, vol. 4, p. 3, 2010.

[VAA 09] VAADIA E., BIRBAUMER N., "Grand challenges of Brain–Computer Interfaces in the years to come", *Frontiers in Neuroscience*, vol. 3, no. 2, pp. 151–154, 2009.

[VID 73] VIDAL J.J., "Toward direct brain–computer communication", *Annual Review of Biophysics and Bioengineering*, vol. 2, no. 1, pp. 157–180, 1973.

[WOL 06] WOLPAW J., LOEB G., ALLISON B. *et al.*, "BCI Meeting 2005–workshop on signals and recording methods", *IEEE Transaction on Neural Systems and Rehabilitation Engineering*, vol. 14, no. 2, pp. 138–141, 2006.

[ZAN 11] ZANDER T., KOTHE C., "Towards passive Brain–Computer Interfaces: applying Brain–Computer Interface technology to human-machine systems in general", *Journal of Neural Engineering*, vol. 8, 2011.

PART 1

Anatomy and Physiology

1

Anatomy of the Nervous System

This chapter's objective is not to describe the nervous system in detail, which would be impossible to do in just a few pages, but rather to provide readers who are interested in Brain–Computer Interfaces but who are not an experts in anatomy, with some basics of neuroanatomy and functional anatomy as well as the vocabulary used to talk about them. Readers looking for greater depth and precision in the description of anatomical structures may consult reference books in neuroanatomy (we can cite for their clarity and exhaustiveness [KAM 13, CHE 98, DUU 98])

This description seeks to provide a general understanding of the structure of the adult nervous system, its main constituents and their principal functions, and to thereby better understand the pathologies associated with it.

This chapter will first provide a general description of the nervous system, and it will then focus on a description of the central nervous system (CNS), as well as that of the peripheral nervous system (PNS). In the last section, we will succinctly describe the main pathologies that can be addressed through the use of Brain–Computer Interfaces.

Chapter written by Matthieu KANDEL and Maude TOLLET.

1.1. General description of the nervous system

A neuron is composed of a cell body and an axon, which terminates in a synaptic area. The information that travels through it is an electric signal that corresponds to a depolarization of the axonal membrane: the action potential. In this way, the axon transmits the action potential up to the synapse, the area of communication between neurons. Molecules emitted at the synapses under the influence of action potentials are called neurotransmitters. These neurotransmitters may either be excitatory or inhibitory and thus determine the response obtained.

Neurons are organized in pathways, tracts or networks whose connections determine their roles. Traditionally, a distinction is made between the CNS and the PNS. It is common to talk about efferent neurons, which transmit information from the CNS to the PNS, and afferent neurons, which transmit information from the PNS to the CNS.

The CNS includes the encephalon, which is enclosed in the skull, and the spinal cord in the spinal canal. The encephalon is itself composed of the brain stem, the cerebellum and the two hemispheres of the brain. The brain stem, located in the most caudal part of the encephalon, gives way to 12 pairs of nerves that are known as cranial nerves. The cerebellum is located in the back of the brain stem. Each hemisphere is composed of several lobes (frontal, parietal, temporal, occipital and the insular cortex). From a functional perspective, each hemisphere has its own specific functions, especially for the most complex functions (for example language in the frontal and temporal areas of the dominant hemisphere, spatial orientation in the right parietal lobe, the organization of complex gestures in frontal lobe, etc.).

The cortex, which is located on the surface of the hemispheres, is composed of gray matter that contains neuron cell bodies and is organized into six layers. The basal ganglia are located at the base of the hemispheres. These are also composed of gray matter. White matter contains myelinated axons from CNS neurons and it makes it possible to establish connections between different parts of the CNS through associative fibers (connecting parts of the cortex to each other or to the basal ganglia) and through fibers that stretch out toward the spinal cord.

The spinal cord, which contains ascending fibers and descending fibers, transmits all motor, sensitive and vegetative information between the encephalon and the PNS. It is also composed of gray matter and is the regulation center for a certain number of reflex actions.

The roots that give way to the PNS arise from the spinal cord. These roots form, passing through the (brachial and lumbosacral) plexuses, the entire set of nerve trunks that make it possible to innervate the skeletal muscles (efferent motor fibers) to transmit sensory (sensitive afferent fibers) and vegetative (efferent and afferent vegetative fibers) information.

Different systems (motor, somatosensory, sensory) may have either ascending or descending pathways, going from the peripheral receptor to the area of the brain involved in interpreting the signal, or going from the cortex all the way to the effector (for example the muscle). We may cite, for example, the descending motor tracts distributed in a (corticospinal and corticobulbar) pyramidal pathway, which is the pathway for voluntary motion. We may also cite extrapyramidal pathways, which include other motor pathways. Other pathways include sensitive, visual, auditory, vestibular and olfactory tracts.

1.2. The central nervous system

The CNS includes the encephalon, which is located in the skull, and the spinal cord, which is located in the spinal canal.

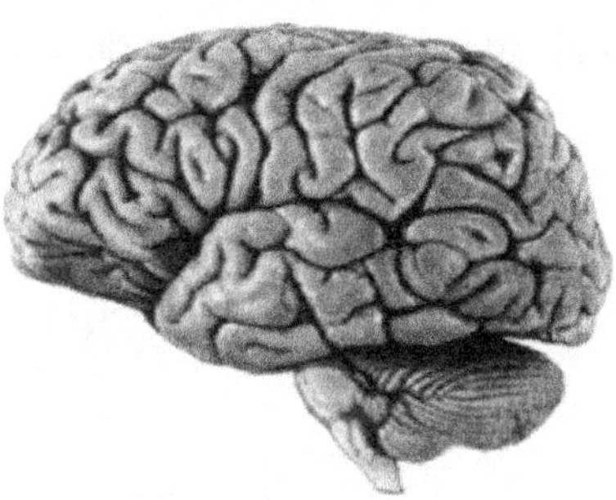

Figure 1.1. *General view of the human encephalon (http://lecerveau.mcgill.ca)*

The encephalon (Figure 1.1) is usually composed of the following structures:

– the telencephalon;

– the diencephalon;

– the brain stem itself comprising the midbrain, the pons and the medulla oblongata. The cerebellum is located in the back of the pons, which is connected to the pons through the cerebellar peduncle.

It is also possible to describe the encephalon from its formation at the embryonic stage. In such a case, we can distinguish between the hindbrain, which will become the medulla oblongata and the metencephalon (pons and cerebellum), the midbrain and the prosencephalon, which will turn into the diencephalon and the telencephalon.

1.2.1. *The telencephalon*

The cerebrum is composed of two hemispheres (right and left) that are connected to one another through white matter tracts (especially by the corpus callosum). The surface of each hemisphere has a folded aspect, which makes it possible to individualize the lobes (Figure 1.2): the frontal lobe, the parietal lobe, the occipital lobe and the temporal lobe on the surface, and the insular lobe on the inside. These lobes are separated by sulci: the central sulcus (also known as the fissure of Rolando), the lateral sulcus or Sylvian fissure, the parietooccipital sulcus and the temporal-occipital sulcus.

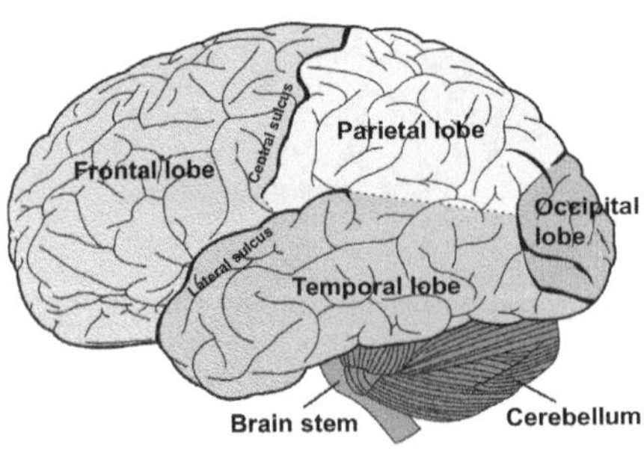

Figure 1.2. *General view of the cortex's surface, main lobes and sulci*

The surface of each lobe itself includes several convolutions, which are known as gyri, and which make it possible to individualize the most superficial parts of the cortex. Despite variations of this structure among different individuals, it is possible to individualize sulci, fissures and gyri in most subjects with relative constancy either in morphological or functional terms.

Korbinian Brodmann, an early 20th Century neurologist and neuropsychologist, established a map of the cerebral cortex by describing 52 areas based on the tissue and histological composition of the cortex (cytoarchitectonic analysis). These are known as Brodmann areas. Brodmann attributed a specific function to each of them. Some of those areas are now subdivided into subareas, and that mapping is still used today [GAR 06].

The functional role of the different areas of the cerebral cortex is traditionally described in the following manner:

– the primary areas, which include the primary motor cortex, and areas that receive sensory stimuli: primary somatosensory cortex (parietal lobe) for sensory information, primary auditive cortex (temporal lobe) and primary visual lobe (occipital lobe);

– the secondary areas, which correspond to elaborate information processing that may be plurimodal, and associative areas, whose functions are more amodal (cognitive and attentional functions) and that most notably make it possible to pay attention to stimuli to identify them. Cognitive functions are processed in such areas.

Let us now review the different lobes:

– *The frontal lobe*: The frontal lobe is composed of the precentral gyrus, the premotor areas and the prefrontal areas. In the dominant hemisphere, it contains Broca's area, which is considered the area of speech production. It is delimited by the central sulcus, which separates it from the parietal lobe, and by the lateral sulcus, which separates it from the temporal lobe.

The primary somatomotor cortex (Brodmann area 4, often called M1), which is located on the precentral gyrus, controls voluntary motor activity. Its efferent fibers form the main part of the pyramidal tract, responsible for direct motion. To every point on the precentral gyrus corresponds a part of the body that it controls: this is called functional somatotopy. To illustrate this, a map known as the cortical homunculus has been created [PEN 50] (Figure 1.3).

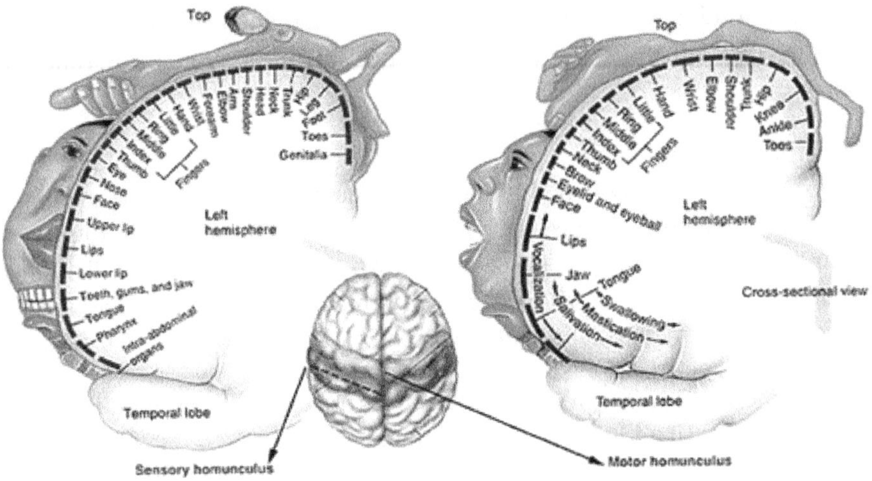

Figure 1.3. *Representation of the motor and sensory homunculi (http://www.corpshumain.ca)*

A lesion in this area can lead to a large or small contralateral paralysis, which corresponds to the projection described in the motor homunculus (hemiparesis).

The prefrontal cortex plays an essential role in determining behavior, motivation and organizational planning, and decision execution capacities. Especially developed in humans, the prefrontal cortex plays an important role in thought elaboration and personality development. Along with the basal ganglia, it is involved in complex motor learning and contributes to long-term memory. In neuropsychology, it is common to speak of executive functions, which include all so-called higher level cognitive functions. Damage to the prefrontal cortex can bring about motor skills learning disorders (for complex tasks), as well as behavioral disorders (lack of initiative, disinhibition, difficulty in planning simple or complex tasks, etc.).

The premotor cortex (which includes the lateral premotor cortex on the outer layer of the frontal lobe, and the supplementary motor area on the midline surface of the hemispheres) is located just anterior to the primary motor cortex.

It is the site of movement planning and organization tasks. Several association fibers connect it to the motor cortex, the cerebellum, the thalamus and the basal ganglia. It makes it possible to select the appropriate movements needed to carry out a desired action. A lesion in the premotor cortex can compromise the capacity to carry out movement toward a specific goal. This is known in clinical terms as dyspraxia.

Broca's area is the area that controls speech production. An injury in Broca's area, which is most often located on the surface of the left hemisphere, can lead to an expressive aphasia. Patients retain the capacity to understand language, but they omit words or employ non-grammatical syntax when attempting to express themselves.

– *The parietal lobe*: the parietal lobe is located between the central sulcus, the lateral sulcus and the parietooccipital sulcus. It comprises the postcentral gyrus, the superior parietal lobule and the inferior parietal lobule.

The primary somatosensory cortex is located in the postcentral gyrus. It receives sensory information and makes it possible to interpret it (as pain, temperature, touch, discrimination, vibration, relative joint positions, etc.). Similarly to the motor homunculus, the organization of the primary motor cortex gives rise to a sensory homonculus (Figure 1.3, left).

An injury in the parietal lobe can produce several different disorders: attentional disorders such as hemispatial neglect (especially in the left hemisphere), sensory extinction, body image disorders and spatial agnosia.

– *The temporal lobe*: the temporal lobe is located in the lower face of the cerebrum and is bounded by the lateral sulcus and the preoccipital notch, which is poorly defined in anatomical terms. From an architectonic standpoint, it is composed of the transverse temporal gyrus (primary auditory cortex), the associative auditory cortex (including Wernicke's area) and the associative temporal cortex involved in language memory. The information coming from each auditive nerve is bilaterally projected in the the primary auditory cortices, for which it is possible to describe a tonotopy (activated areas are associated with a specific sound frequency).

An injury to the temporal lobe can bring about auditory disorders (cortical deafness, auditory agnosia, auditory hallucinations, transcortical sensory aphasia if the dominant hemisphere is injured, quadrantanopia).

– *The occipital lobe*: as previously described, the occipital lobe is separated from the parietal lobe by the occipitoparietal sulcus and from the temporal lobe by the preoccipital notch, although less markedly so.

It is divided into three occipital gyri: the cuneus, which is separated from the lingual gyrus by the anterior calcarine sulcus and the occipitotemporal gyrus, which is separated from the lingual gyrus by the collateral sulcus.

The primary visual cortex (V1) and the associative visual cortex are located at the level of the occipital lobe. Much like for the primary motor cortex, or the primary somatosensory cortex, there is a systematic correspondence, with every point in our visual field being represented in a specific area of V1. This property relating V1 to the retina is known as retinotopy.

– *The insular cortex or lobe*: the insular cortex is the "5th lobe" in our brain. It is located deep within the frontal, parietal and temporal opercula. Although it is less precisely understood, we know that the insular cortex is involved in corporal self-consciousness, in pain intensity recognition and, for example, in the perception of heartbeat frequency.

1.2.2. *The diencephalon*

This is the part of the brain that connects the brain stem to the cerebral hemispheres. It bounds the third ventricle on both sides. The diencephalon is composed of the thalamus, the hypothalamus (center of the vegetative nervous system), the subthalamus, the epithalamus and the basal ganglia. It is bounded on the side by the internal capsule.

The thalamus is a set of ganglia that receive a variety of afferents. It is the true relay center of sensory, motor and emotional information for the cerebral cortex. It is divided by the medullary laminae into medial, lateral, anterior and posterior nuclei. Each area is a specific anatomical relay that determines its functions.

The thalamus is in part controlled by thalamic reticular formation, a set of neurons that separate it from the internal capsule, determining the state of cortical activity.

The thalamus is involved in the transmission of visual, auditory, somatosensory and vestibular information. It plays a role in awareness of transmitted pulses (perception of pain or heat, and of emotional perception) and contributes to the execution of movement.

It receives several afferents (hypothalamus, mammillary body, basal ganglia, sensory afferents, cerebellum, etc.) and it emits several efferents (prefrontal cortex, cingulate gyrus, associative parietal cortex, pallidum, auditory and visual cortex, motor cortex, etc.). It is in this sense a true circuit board for all information processed and transmitted by the nervous system.

The hypothalamus, which is the center of all vegetative functions, is divided into anterior, tuberal and posterior regions. It is located above the thalamus and is connected to the pituitary gland by the pituitary stalk. Because of its multiple connections, it can control the vegetative system (through the sympathetic and parasympathetic nervous systems). It can also regulate homeostasis (maintaining constant internal conditions, including temperature and electrolyte equilibrium). The hypothalamus is also involved in food intake and cardiac cycle regulation.

The pituitary gland plays a role in endocrine function regulation. The epithalamus contains the epiphysis, an endocrine gland that produces melatonin and contributes to sleep cycle and circadian cycle regulation.

The basal ganglia are part of the extrapyramidal system, and include the putamen, the caudate nucleus and the globus pallidus. The putamen and the caudate nucleus form the neostriatum, while the globus pallidus and the putamen form the lenticular nucleus.

The striatum receives not only a large amount of information coming from motor brain areas, but also from the temporal, parietal and limbic areas. It plays a major role in the execution of voluntary movement carried out in a regular and fluent manner, as well as in determining behavioral and cognitive functions. The striatum is, among others, connected to the thalamus and to the substantia nigra (located in the midbrain) and is also involved in involuntary movement control.

From an anatomical standpoint, the caudate nucleus receives information from association areas and from motor areas of the frontal lobe, which most prominently allow for control of complex movements.

The putamen receives information coming from the primary and secondary somatosensory areas of the parietal lobe, the secondary visual areas, the associative auditory areas of the temporal lobe and the frontal lobe's motor and premotor cortices. Its neurons are activated when a given movement is about to occur. The striatum is an indispensable element in movement activation.

Within the basal ganglia, it is common to talk about the subthalamic nucleus, also known as corpus Luysi or Luys' body. This anatomical structure is particularly interesting in the scope of clinical practice because the neurosurgical implantation of stimulation electrodes in this site is an effective treatment for Parkinson's disease.

The substantia nigra, located further down in the midbrain, is also traditionally considered a part of the basal ganglia.

1.2.3. *The brain stem*

The brain stem is composed of the midbrain, the pons and the medulla oblongata (Figure 1.4). It is located at the back of the cerebral hemispheres and extends as far back as the foramen magnum. The nuclei of the cranial nerves emanate from the gray matter of the brain stem at different levels. The (I) olfactory and (II) optic nerves are expansions of the CNS. All ascendant and descendant pathways (pyramid, spinothalamic tract, posterior column, medial lemniscus pathway, etc.) pass through the brain stem.

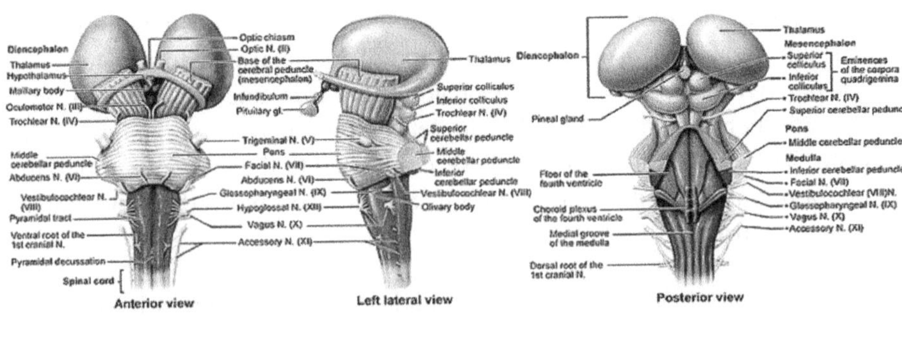

Figure 1.4. *Brain stem, anterior, lateral and front views (http://www.corpshumain.ca/)*

The midbrain connects the diencephalon to the pons. It gives way to the (III) oculomotor and (IV) trochlear nerves, which contribute to oculomotricity. Among others, the nuclei of the substantia nigra, red nuclei (extrapyramidal system) and the reticular formation are found in the midbrain.

The pons (also called pons Varolii) connects the midbrain to the medulla oblongata and is itself connected to the cerebellum through the cerebellar peduncles, which are crossed by the trigeminal nerve (V, transmitting sensation in the face, a part of the tongue and muscles responsible for mastication). The trochlear (VI, involved in oculomotricity with III and IV), facial (VII, whose main function is face and muscle movement) and vestibulocochlear (VIII, hearing and balance organs) nerves arise from it.

Finally, the medulla oblongata or bulb extends from the pons all the way to the first cervical nerves. On the ventral part, we find the corticospinal fibers (pyramidal pathways). This is the anatomical site of the intersection or decussation of those fibers, which is responsible for the left side of the brain controlling the right side of the body and vice versa.

From the medulla oblongata arise the glossopharyngeal nerves (IX) and the vagus nerve (X) (IX and X are involved in the oropharynx's locomotion and sensory capacities, which are responsible for swallowing and phonation), as well as the accessory nerve (XI, motor innervation for some muscles in the cephalic region) and the hypoglossal nerve (XII, tongue motor innervation).

The brain stem contains centers included in the autonomic nervous system (ANS). In this way, the parasympathetic nervous system's efferents originate at the brain stem at nuclei III, VII, IX and X of the cranial pairs. It is a cholinergic system that is controlled by the hypothalamus (superior vegetative nervous system) described above. This system controls the functions of several organs (the iris, lacrimal and salivary glands, the heart, lungs, stomach, pancreas, intestinal smooth muscle tissue and digestive glands), and thus confers a vital function to the brain stem.

The brain stem is the site of several reflexive and vital activities.

1.3. The cerebellum

The cerebellum, which is shown in Figure 1.5, is a fundamental part of the CNS, as the major coordination center for equilibrium and muscle tone. It enables the completion of specific motor tasks.

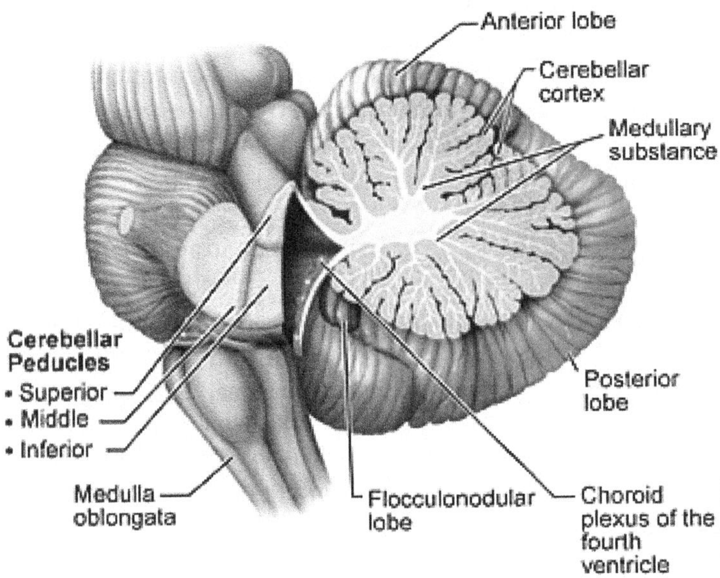

Figure 1.5. *Brain stem and cerebellum (http://www.corpshumain.ca), lateral view*

At the anatomical level, it is composed of two lateral hemispheres (right and left) and the cerebellar vermis in the medial zone. It is connected to the dorsal side of the brain stem by three pairs of cerebellar peduncles and is covered by the tentorium cerebelli, which separates it from the cerebral hemispheres. It is divided into ventral, superior and inferior faces. It receives inflows coming from the cerebral cortex, the brain stem and the medulla oblongata, thus constituting several afferent and efferent connections.

From a functional perspective, the cerebellum is divided into three parts determined by their phylogenetic origin: the neocerebellum, the paleocerebellum and the archicerebellum.

The neocerebellum is composed of two cerebellar hemispheres. It is the largest of the three parts. It receives information coming from the cerebral cortex through the corticopontocerebellar pathway, and information coming from the oliva through the olivopontocerebellar pathway.

The neocerebellum is the fundamental element that makes it possible to carry out voluntary and involuntary movement in a precise manner. An injury to this part of the brain would cause motor disorders, affecting coordination in particular.

The paleocerebellum is composed of the ventral part of the cerebellum. It receives information from the spinal cord through the dorsal and ventral spinocerebellar tracts, as well as the cuneocerebellar tract. Its role is very important in the maintenance of equilibrium and involves contributing to muscular tone and agonistic–antagonistic muscle synergy regulation. An injury to this structure leads to hypotonia.

The archicerebellum is composed of the nodulus and the flocculus. It is connected to the vestibular system and receives information about the head's position in space during movement. It makes it possible to maintain equilibrium and to have appropriate ocular and cephalic reactions to movement. When injured, patients present a so-called ataxic walk (stumbling, with an enlarged support polygon).

1.4. The spinal cord and its roots

The spinal cord is the part of the CNS that goes from the foramen magnum to the skull base, advancing through the spinal canal formed by the vertebrae up until dorsal vertebra L2.

It forms a cylinder with a diameter of approximately one-third of an inch and a length of 15–18 inches in adults (Figure 1.6). Spinal nerves stem from the vertebral column at different heights (eight cervical spinal nerves, labeled C1–C8; 12 thoracic nerves, labeled T1–T12, five lumbar nerves L1–L5, five sacral nerves S1–S5 and one coccygeal nerve).

The roots exit the spinal canal through foramina located between two vertebrae. The whole cord is covered by the meninges membranes (dura mater, arachnoid mater, pia mater) and immersed in cerebrospinal fluid.

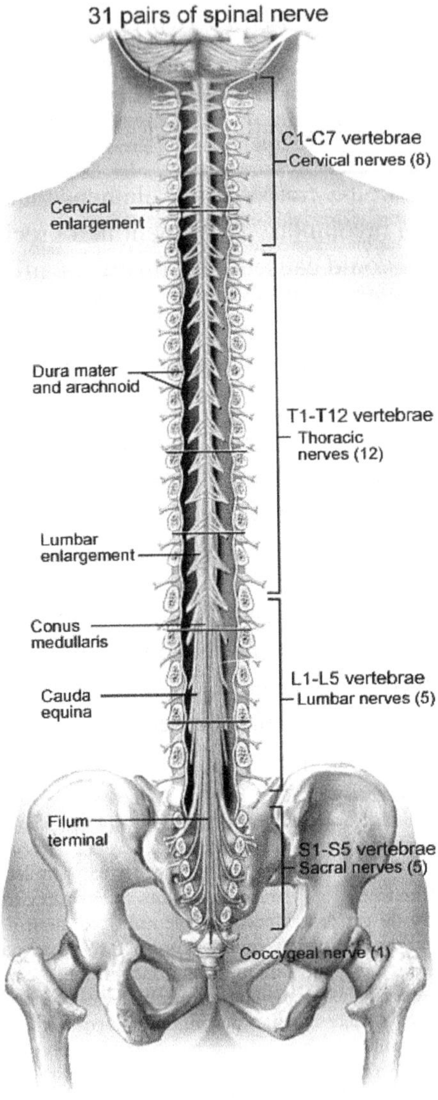

Figure 1.6. *Spinal cord and spinal nerves, general view (http://www.corpshumain.ca)*

During development, growth in the size of the spinal canal is more important than growth of the spinal cord, which is why the lower end of the spinal cord (called conus medullaris) is usually found at L2 in adults. The cauda equina designates the set of nerve roots located between the conus medullaris and the last vertebra.

There are two regions where the spinal cord enlarges, one at the cervical level and another at the lumbar level, and they serve to innervate the upper and lower limbs.

The spinal cord has vegetative fibers and is the center of reflexive motor activity that has the purpose of regulating muscle tone. Surrounding the spinal cord is white matter containing all ascending and descending fibers that transport sensory afferent messages and efferent motor messages. At the center of the spinal cord, there is gray matter, which contains neuronal nuclei.

From an anatomical point of view, it is possible to describe three white matter cords in each half of the spinal cord (Figure 1.7), which contains ascending and descending pathways:

– the posterior cord (cuneiform tract of the upper limb and gracile tract of the lower limb) contains the large diameter myelinated ascending fibers (also known as the posterior column-medial lemniscus pathway), which transmits limbs' conscious position sensibility (proprioception), tact and pressure;

– the lateral cord contains the pyramidal tract and the spinothalamic lateral tract (pain, temperature). The spinocerebellar (dorsal and ventral) tracts transmit unconscious deep sensory information (which most notably makes it possible to regulate axial muscle tone);

– The anterior cord contains ascending spinothalamic ventral fibers (sensitivity to pressure, called protopathic).

Gray matter is usually composed of the central area, the anterior gray column and the posterior gray column. The posterior gray column includes the synaptic connections for all sensory information coming from the dorsal root of the spinal nerve. The anterior gray column contains the bodies of α motor neurons, and gives way to the anterior root of the spinal nerve.

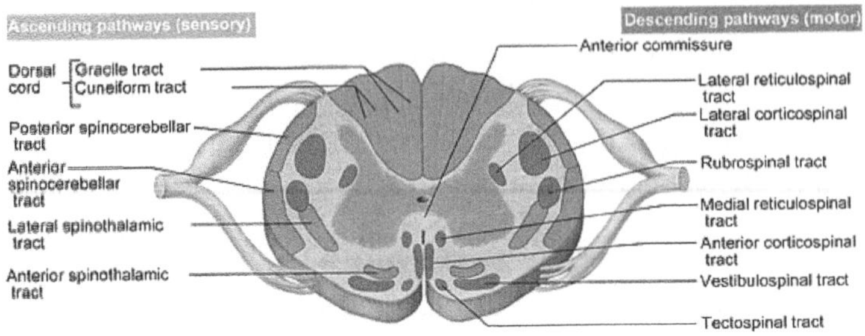

Figure 1.7. *Cross-section of the spinal cord (http://www.corpshumain.ca), diagrammatic representation of ascending and descending tracts*

1.5. The peripheral nervous system

1.5.1. *Nerves*

A nerve is a collection of sensory, motor and/or vegetative axons. When it is exclusively composed of motor or sensory axons, it is called a motor or sensory nerve. When it is composed of a combination of both, it is called a mixed nerve. Its neural fibers are either myelinated or non-myelinated. Myelin is a substance that determines the speed at which nervous influx is conducted. Myelin is produced by Schwann cells.

Sensory fibers are divided into several groups according to their size:

– *Group I*: it consists of type Ia fibers (approximately 17μm, with a conduction speed of 70–120 m/s), which are myelinated and transmit unconscious proprioceptive sensibility through neuromuscular spindles, and Ib fibers (approximately 16μm, with a conduction speed of 70–100 m/s) based on the Golgi apparatus. They transfer information to the cortex through the spinocerebellar tract. Ia fibers also communicate with the spinal cord in the monosynaptic reflex arc;

– *Group II*: aβ and Aγ fibers (8 μm, 15–40 m/s) are myelinated and transmit epicritic sensitivity as well as deep sensitivity;

– *Group III*: a δ fibers (3 μm, 5–15 m/s) are myelinated and they enable the transmission of thermal sensitivity, pain and pressure;

– *Group IV*: C fibers (0.2–1μm, 0.2–2 m/s) are unmyelinated and are responsible for transmission of pain and temperature, as well as coarse tactile sensations.

Motor fibers are divided into α motor neurons, which innervate the striated skeletal muscles that are responsible for movement, and γ motor neurons, which innervate the muscular part of neural spindles, which are key in controlling muscle tone.

Vegetative fibers are described in section 1.5.3, which deals with the ANS.

Neural fibers (myelinated or unmyelinated motor/sensitive/vegetative fibers) are enclosed in a connective tissue called endoneurium. They are first bundled into nerve fascicles enclosed in perineurium, and then several fascicles are enclosed in endoneurium, giving way to a nerve.

1.5.2. *General organization of the PNS*

The PNS is composed of the following:

– efferent motor fibers consisting of a cellular body that arises from the spinal cord's anterior gray column, an axon (which constitutes the motoneuron) and a neuromuscular junction (NMJ), where the axon and the innervated muscle interact. Motor fibers come together to form the ventral root of a spinal nerve;

– afferent sensory fibers transmit peripheral information to the CNS, whose cell body is located in the dorsal root ganglion. They communicate with the spinothalamic and spinocerebellar tracts as well as with the posterior column–medial lemniscus pathway at the level of the spinal cord's posterior horn. They transmit information from the skin, muscles, joints and organs. They form the spinal cord's dorsal root.

Dorsal and ventral roots unite at each metamerism of the spinal nerve (which is composed of motor and sensory fibers that are coupled with vegetative fibers). In total, there are 31 pairs of spinal nerves and 12 pairs of cranial nerves. Spinal nerves are each composed of a ventral and a dorsal root. These bind to the roots of other spinal nerves to form plexuses, which give way to muscular, skin and organ nerves.

We may thus distinguish between the following:

– the *cervical plexus* that is formed by the ventral roots going from C1 (first cervical root) to C4 and innervating the nape and neck muscles. The phrenic nerve (responsible for diaphragm motricity) also originates in the cervical plexus;

– the *brachial plexus* that is formed by the union of ventral fibers going from C5 to T1 (first thoracic or dorsal root). The auxiliary, musculocutaneous, radial, median and ulnar nerves (which are responsible for the upper limbs) in the brachial plexus;

– the *lumbosacral plexus* that is composed of the roots of L1 to S3. The femoral, obturator, iliohypogastric, genitofemoral, ilioinguinal and sciatic nerves, as well as the lateral cutaneous nerve of the thigh, which are responsible for sensory and motor innervation of the lower limbs, originate in the lumbosacral plexus;

– the *pudendal plexus* that is formed by the S2–S4 roots coming together in the pudendal nerve.

It is possible to produce a map of cutaneous areas according to their radicular innervation, known as dermatomes. Each muscle is innervated by a nerve originating in one or another of the aforementioned plexuses. One root innervates a set of muscles, which is known as a myotome.

The junction between a motor neuron axon and a muscle is called the NMJ. A motor neuron's axon splits in two to innervate a certain number of muscular fibers. The set of muscular fibers innervated by a given axon constitutes a motor unit. This system is cholinergic: an action potential leads to the release of acetylcholine within a synapse. Reception of acetylcholine by the postsynaptic nicotinic receptors of the motor endplate leads to the contraction of muscular fibers.

1.5.3. *The autonomic nervous system*

The peripheral vegetative or ANS works involuntarily. It is composed of the sympathetic and parasympathetic systems.

Vegetative fibers innervate the viscera, the glands and the vessels (smooth muscle). They allow for involuntary movement, and they enable the transmission of pain from the organs to the cortex.

The most important functions that are regulated by the ANS are body temperature, sweat, pilomotor functions, the cardiovascular system, breathing, the digestive system, detrusor–sphincter and genitosexual functions.

1.6. Some syndromes and pathologies targeted by Brain–Computer Interfaces

1.6.1. *Motor syndromes*

1) *Peripheral neuropathic syndrome*: this refers to symptoms observed when there is damage to peripheral nerves, regardless of the cause (trauma, intoxication or illness). The symptoms are either localized (injury to the nerve trunk or to a root) or diffused. Flaccid paralysis is the most common symptom, along with amyotrophy and hypotonia. There is also autonomic dysfunction (vasomotor dysfunction, problems with blood pressure regulation, diarrhea or constipation, impotence, etc.);

2) *Pyramidal syndrome:* pyramidal syndrome results from injury to the corticospinal tracts. From a clinical standpoint, it is possible to observe motor command disorders (paralysis) ranging from partial deficit to total deficit. This kind of paralysis is associated with muscle tone disorders and spasticity, which has been described as an overly augmented stretch reflex. Beyond paralysis, spasticity can have major functional consequences, which may involve muscular retractions and joint stiffness. Clinicians look out for this syndrome by studying tendon reflexes that may be overactive;

3) *Cerebellar syndrome*: cerebellar syndrome is the result of either an injury to the cerebellum itself or to one of the cerebellar pathways. Its most common symptom is cerebellar ataxia, which includes gait abnormality and orthostatic intolerance. Patients walk in a "wobbly" manner, and their support polygon is expanded. Muscle tone is diminished (hypotonia) and coordination and voluntary movements are altered. It is possible to observe, for example, dysmetria (lack of coordination in space), adiadochokinesia (lack of coordination in time) and shakiness when executing movements. Speech is also affected, which is known as cerebellar dysarthria;

4) *Extrapyramidal syndrome*: this syndrome is the product of damage to the extrapyramidal system described above. It is associated with hypertonia (called extrapyramidal hypertonia), at-rest shakiness or hypokinesia (difficulty initiating movement and decreased bodily movement). From a clinical standpoint, it is possible to observe postural disorders as well as gait abnormality (trampling or festination, anteflexion, *marche à petits pas*, absence of parachute reaction), dysarthria, micrography (which makes writing illegible) and cognitive disorders (psychic slowness, perseveration, etc.).

1.6.2. *Some pathologies that may be treated with BCIs*

1) *Spinal cord injuries:* spinal cord injuries are most often the product of trauma. A distinction is made between instances where the damage occurs at the site of the lesion, in which case peripheral neuropathic syndrome can be observed, and instances where the damage produces a pyramidal syndrome, or sensory, vesicosphincteric and genito-sexual disorders. Clinically, it is common to describe the level of the spinal injury in terms of the lesional syndrome (last healthy metamerism).

Spinal cord injuries are divided into complete (all motor and sensory functions are affected) and incomplete injuries. Tetraplegia refers to injuries affecting all four limbs, and paraplegia refers to injuries affecting only the lower limbs.

Spinal cord injuries are classified according to the location of the affected area. For example;

- Brown–Séquard syndrome or hemiparaplegic syndrome: the spinal injury is unilateral, with paralysis on the side of the body where the lesion occurred, as well as damage to proprioception. Contralateral thermoalogic damage is observed,

- anterior compartment syndrome: motor and thermoalgic deficit without any additional sensory damage,

- posterior cord syndrome: purely sensory damage and pain is abundant; the sensory deficit concerns proprioception, but preserves thermoalgic sensation;

2) *Locked-in syndrome*: locked-in syndrome (LIS) is also known as cerebromedullospinal disconnection or de-efferented state: it is the result of an

injury to the motor pathways at the brain stem, most often on the protuberance. Voluntary limb and face movement is affected, as are deglutition, phonation and oculomotricity. Sensory and cognitive functions are usually preserved. The patient is therefore capable of perceiving his or her environment and remains perfectly conscious. The capacity to raise the eyes is often retained, making it possible for patients to communicate. One of the most important challenges when caring for patients affected with LIS is establishing effective communication.

3) *Hemiplegia*: hemiplegia is characterized by a bodily motor deficit on the opposite side of a lesion to the primary motor cortex or of the subcortical pyramidal tract. The deficit can either be total or partial (in which case it is referred to as a hemiparesis). Hemiplegia is most often the product of a cerebrovascular accident or stroke and produces other symptoms according to the affected area. Related cognitive disorders can make access to BCIs difficult for patients.

4) *Amyotrophic lateral sclerosis (ALS) also known as Lou Gehrig's disease or Charcot disease*: ALS is a form of degenerative, progressive damage to both central motor neurons and peripheral motor neurons. It is therefore associated with pyramidal syndrome and peripheral neuropathic syndrome. Patients first display distally with atrophy, fasciculations and quick reflexes, as well as spasticity. There is no sensory damage.

The disorder affects the limbs, as well as deglutition and phonation, making it difficult for patients to communicate.

1.7. Conclusions

It is difficult to describe neuroanatomy with any degree of precision in just a few pages. This chapter's authors hope that readers who are not already familiarized with anatomy will have understood the general structure of the nervous system and will have acquired its most important terms, thereby making it possible to understand the following chapters. The description provided here is partial by design, and has therefore left out important structures (such as the ventricular system), with an important role in pathology, but which have been deemed to fall outside the scope of this book. This being said, it is important for readers interested in a specific structure of the nervous system to consult reference works in neuroanatomy. Indeed, knowledge of neuroanatomy is necessary in order to understand how

Brain–Computer Interfaces work, and sharing common neuroanatomical knowledge is important for the fruitful collaboration of actors from different fields.

1.8. Bibliography

[CHE 98] CHEVALLIER J.M., *Anatomie, Neuro-anatomie*, Médecine-Sciences, Flammarion, 1998.

[DUU 98] DUUS P., *Diagnostic neurologique, les bases anatomiques*, DeBoeck Université, Bruxelles, 1998.

[GAR 06] GAREY L., *Brodmann's Localisation in the Cerebral Cortex*, Springer, New York, 2006.

[KAM 13] KAMINA P., *Anatomie clinique, Tome 5, Neuroanatomie*, Maloine, Paris, 2013.

[PEN 50] PENFIELD W., RASMUSSEN T., *The Cerebral Cortex of Man: A Clinical Study of Localization of Function*, MacMillan, New York, 1950.

2

Functional Neuroimaging

Neuroimaging has experienced significant progress in the past few decades, especially since the advent of magnetic resonance imaging (MRI) in the 1970s and 1980s and functional MRI (fMRI) at the beginning of the 1990s [OGA 90].

There are two types of neuroimaging. The first type is structural imaging, which employs magnetic resonance. Images thus produced determine the brain's anatomy as well as the white matter tracts that connect the different regions of the brain (this method is known as "tractography"). The second type is functional imaging, which has the objective of identifying regions involved in a given cognitive task or in certain pathologies, as well as understanding the interactions between the different regions.

Several different methods are used to map the regions of the brain, each of which employs very different mechanisms. The activation of a given brain region produces a host of events, including electromagnetic fields and changes in blood flow, which provide signals gathered by the different mechanisms.

In this chapter, we will provide an introduction to the functional neuroimaging methods that can have implications for brain–computer interfaces and related fields. We will first study fMRI, then electroencephalography (EEG), magnetoencephalography (MEG) and finally simultaneous EEG-fMRI. EEG and MEG are establishing themselves as

---

Chapter written by Christian BÉNAR.

fully-fledged neuroimaging systems, since they now produce activation images on the cortical surface thanks to inverse problem methods. We will also outline the general principles guiding signal processing techniques that make it possible to detect brain activity.

2.1. Functional MRI

2.1.1. *Basic principles of MRI*

MRI provides a signal proportional to the number of a certain kind of atoms at each point in the brain. In the context of biology, hydrogen is commonly studied, since it is present in large quantities due to the presence of water in tissues.

Magnetic resonance is based on the fact that an atom placed in a strong B_0 magnetic field has a spin that aligns with the field [HOR 96]. We can represent this metaphorically by saying that each atom behaves like a spinning top, revolving on its own axis which in turn also revolves in the manner of a cone of precession (just like a spinning top that slows down). The rotation frequency depends on the magnetic field strength and on the nature of the atom, which is known as its Larmor frequency. When a magnetic wave at the Larmor frequency is sent to the atom, its magnetic moment changes dramatically and then returns to normal, while emitting a signal at the same frequency as that of the wave that excited it – this is magnetic resonance.

The key element of this technique is to slightly modify the magnetic field in space using gradient coils, which allows encoding space with frequency: at each point in the brain, the resonance frequency is different. Brain images are thus obtained using the inverse Fourier transform.

Each volume element is known as a "voxel". MRI has long used a B_0 field in the order of 1.5T, and now often uses 3T or even 7T, which makes it possible to obtain higher resolution and/or a higher signal to noise ratio.

2.1.2. *Principles of fMRI*

At the beginning of the 1990s, the Ogawa team demonstrated that MRI can detect the presence of deoxyhaemoglobin (dHb), which modifies the magnetic field locally by causing a phase shift in oxygen atoms, thus diminishing the gathered signal at that point. This is known as the "blood oxygen level dependent" (BOLD) effect [OGA 90]. When a region of the brain is activated, oxygen is consumed and dHb is produced, giving way to a temporary drop in MRI signals. However, the vascular response quickly supplies blood to the activated region, diluting dHb and increasing the BOLD signal [HOG 99]. The hemodynamic signal takes about 5 seconds to reach it peak. By rapidly producing images (usually echo planar imaging) every 2 or 3 seconds, it is possible to obtain a series of functional images that allow tracking of the brain's activity at intervals of several minutes.

When no stimulation protocol is used, other than requesting the patient to remain calm and not to think about anything in particular, this is called resting state imaging, a topic that is currently of great interest to the neuroscience community [RAI 07]. In order to study the brain's response to a given type of stimulation, there are two strategies: either sending sets of same-kind stimuli to obtain a strong response in a given region (block-related design), or sending different kinds of interwoven stimuli (event-related design). Event-related design has several advantages in terms of data analysis [BUR 98].

2.1.3. *Statistical data analysis: the linear model*

fMRI detection of active regions is traditionally accomplished with a linear model (general linear model) [WOR 02]. The model predicts the BOLD response at each point in the brain (at each "voxel") through a linear regression. Regressors considered "of interest" represent the brain's response to stimulation. They are typically obtained using the convolution of pulses representing stimulation times with a model of the hemodynamic response. Regressors "of no interest" model other sources of fluctuation observed in the data, like the subject's movements and low frequencies.

For each combination of regressors of interest (for example the contrast between two conditions), it is possible to produce a statistical map that tests the significance of that regressor (Figure 2.1). Since the test is repeated on all of the brain's voxels (in the order of several million), one of the major difficulties is controlling for the multiple comparisons in order to decide which region is active or not.

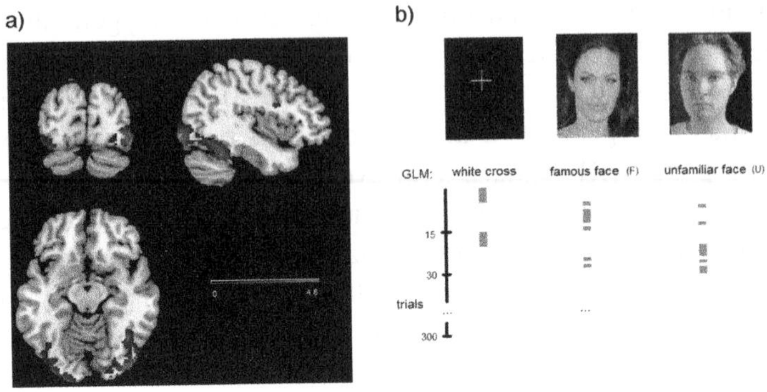

Figure 2.1. *Example of an fMRI statistical analysis in face recognition design (data: Jonathan Wirsich and Maxime Guye, CEMEREM-CRMBM, UMR 7339): a) statistical map contrasting faces with a white cross; b) occurrence time for events used in the linear model (white cross, famous face, unfamiliar face) (from [WIR 14]). For a color version of this figure, see www.iste.co.uk/clerc/interfaces1.zip*

A classical strategy is to control type I errors (false positives), that is to say the probability of exceeding the threshold in the null hypothesis (no activation; i.e. there is just noise at that voxel). Let us imagine that the threshold is set at 5% ($p < 0.05$), which is the norm in parametric statistics. If the test is repeated a large number of times (e.g. several million times), the number of false positives will increase: the probability of exceeding the threshold by chance increases with the number of tests. For a small number of tests, n, it is possible to use a Bonferroni correction, which assumes the tests to be independent and uses a threshold of $p = 0.05/n$. However, in imaging data, there is a strong correlation between neighboring points, and the large number of points would make the threshold far too conservative (it would lead to an overly large number of false negatives).

A large amount of research has been carried out in the field of fMRI statistics. Controlling for false positives can be achieved with Gaussian field theory, which uses the spatial correlation of noise. When this theory is applied to amplitude peaks, it produces relatively high thresholds. Indeed, it controls for the error rate known as the "family wise error rate" (FWER), which represents the probability that a single voxel exceeds the threshold [WOR 96], which is a very tight condition [NIC 03]. Another possibility is to apply it to the size of clusters i.e. the connected voxels exceeding an arbitrary threshold (for example $p = 0.001$ without correction) [FRI 94]. Yet another possibility is the "false discovery rate" (FDR) threshold, which is less demanding since it tolerates a certain percentage of false positives [GEN 02].

One interesting option for determining thresholds is using non-parametric thresholds. For example, when there are two conditions, it is possible to randomly exchange the labels "condition 1" and "condition 2" through stimulation repetitions and to measure the maximum statistic value throughout the brain (thereby controlling the FWER). By repeating this label permutation a large number of times, it is possible to obtain a distribution of the maximum statistic in the null hypothesis (no significant difference between the conditions that have been randomly permuted) [NIC 02]. On this distribution, it is possible to calculate the threshold corresponding to a given critical value, for example $p < 0.05$.

2.1.4. *Independent component analysis*

Independent component analysis (ICA) is a multivariate method that makes it possible to blindly extract the structure present in the data [COM 94]. Each component has a spatial map, which represents the voxels whose activity evolves in a similar manner, and an associated temporal part; that is, the fluctuations in time of that component's amplitude. The observed data are the sum of all components. Since this linear decomposition is non-unique, it is necessary to apply a constraint, which in this case is the statistical independence of spatial maps [MCK 98]. The formula for ICA is the following: $U = W X$, where X represents fMRI data (time x voxels), W is an "unmixing" matrix (components $\times$ time) and U is the matrix of spatial components (components $\times$ voxels). ICA assumes the lines of U to be mutually independent (that is, according to the spatial dimension).

This type of analysis has produced excellent results in resting state imaging and has made it possible to extract several activity patterns that are reproducible across subjects [DAM 06]. Another reason this method is interesting is because it makes it possible to eliminate physiological noise from the data [PER 07].

2.1.5. *Connectivity measures*

The linear allows identification of active regions during a given protocol. Connectivity measures seek to define the interactions between those regions. In other words, they seek to determine "who talks to whom".

The simplest method is correlation (linear or nonlinear). For this measurement, it is important to take into account physiological fluctuations such as breathing and heartbeat, which produce correlations in the signals that are not directly related to brain activity.

Typically, correlation is computed between each pair of voxels or cerebral regions. It has been used in several contexts: cognitive tasks, at resting state or pathologies such as epilepsy [BET 11].

This type of bivariate measurement is very useful, but it is also sensitive to indirect correlations. For example, if region A is related to region B, and region B is related to region C, a correlation may be found between A and C that does not correspond to a physical link between those two regions. In order to eliminate indirect links, several methods have been proposed. They may be based on correlation such as partial correlation [MAR 06], or on a regression through multivariate linear models (combined with a Granger causality test). They may also employ *a priori* assumptions, like structural equation modeling, or dynamic causal modeling [PEN 04].

Granger causality is of great interest to the neuroimaging community. By definition, there is causality as defined by Granger in a signal X toward a signal Y if X's past state helps predict the present of Y better than the past state of Y alone [ROE 05]. It is important to note that the term "causality" is controversial, since in order to measure causality in a physical system, it is usually necessary to modify the system (which could, for example, be done by actively inhibiting a given region). Some researchers highlight the fact that Granger causality is strongly related to correlation [DAV 13], and that it is difficult to extract time information by using fMRI responses [SMI 11].

2.2. Electrophysiology: EEG and MEG

2.2.1. *Basic principles of signal generation*

The basic element generating a signal is a neuron. When it receives a postsynaptic potential, a movement of ions is produced throughout the membrane, which creates a current that propagates in the head [LOP 05]. The current generated by a single neuron (a small dipole with an elementary current) is undetectable. In order for there to be signals with an amplitude sufficiently large to be measured, a large number of neurons must be active at the same time, which can be represented with an equivalent dipole. A spatial organization producing a summation of currents is also necessary, which is the case in cortical pyramidal neurons, which are aligned in parallel (see Chapter 3). EEG is the measure of the electric potential difference between a point on the scalp and a reference electrode. This difference in potentials is created by currents that propagate in the head. A major factor in current propagation is the skull, which is less conductive than the brain and the scalp. The skull is a place where currents attenuate and diffuse, which produces a spatial smoothing of electric potentials [NUN 05].

MEG is a measure of the magnetic field on the surface of the head (though not necessarily in contact with it). The propagation of a magnetic field is much less influenced by the media through which it travels than that of an electric field. MEG is therefore relatively insensitive to the skull's presence, which produces good spatial properties. On the other hand, a so-called "radial" dipole – that is one pointing toward the surface – produces only a very weak magnetic field [HÄM 93].

2.2.2. *Event-related potentials and fields*

The simplest analysis of EEG and MEG is carried out at the level of each sensor. On continuous data, it is possible to characterize the frequency content using spectral analysis. For a stimulation or cognitive protocol, the average response to each condition (with a condition being a type of stimulation or a cognitive task) is separately calculated [COL 96].

This is referred to as event-related potentials for EEG and event-related fields for MEG. A statistical test is carried out in order to compare responses at each point in time and at each sensor, and to detect the regions/temporal neighborhoods where a difference is induced by the protocol (see section 2.2.7).

In general, it is necessary to calculate averages over a large number of repetitions of the same stimulus or of the same task because a single response is too weak and too easily fades into noise. It is important to note that averaging may mask some variability at the level of individual events. For example, temporal variability (i.e. jitter) that may be more pronounced in one condition can result in a lower amplitude on average: a latency effect becomes an amplitude effect [HOL 06]. Noise reduction methods can be used to estimate responses at each individual event, and thus overcome the limitations inherent to averaging.

2.2.3. *Source localization*

MEG and EEG data are measured on the surface of the head. It is therefore useful to employ source localization techniques to estimate the location of that activity within the cerebral cortex [BAI 01]. The challenge is to calculate the electric and magnetic fields produced by an elementary dipole current. It requires a model of the head and of the different tissues' conductivity. The simplest model is a sphere, which is the first to have been used. More recently, finite element surface and volume models have been developed in order to have a more accurate representation of the tissues (brain, skull, scalp). The inverse problem uses the direct problem to infer the amplitude and location of sources within the brain that have produced the measurements. This problem is ill-defined, since it has an infinite number of possible solutions and is very sensitive to noise. It is therefore necessary to limit the problem by making mathematical assumptions. Several approaches have been proposed to solve the problem, which may differ in terms of the constraints they impose, but which often have strong connections to one another [BAI 01].

Most approaches are based on a linear model that describes observed fields as the superposition of the activity of one or several dipolar sources. When a small number of dipolar sources is used, the procedure is known as an equivalent dipole solution [SCH 85]. A large number of sources uniformly distributed in the volume or on the cortex is referred to as distributed sources. Since in that case there are many more unknowns than measurements, a regularization method must be used, the most common one being the minimum energy constraint on the sources [HÄM 94].

For sources distributed along the cortex, an additional constraint is used, assuming that their dipoles are perpendicularly oriented toward the cortical surface, which is consistent with the hypothesis about the importance of pyramidal cells in signal generation. Another approach (known as "beamforming") uses spatial filters, which seek to estimate activity at a given point of the brain by limiting the influence of other regions [VAN 88].

All these methods make it possible to observe brain activity with high temporal resolution (in the order of milliseconds). Figure 2.2 illustrates a source localization result on MEG over an average of eight epileptic discharges considered at the discharge peak. The temporal progression is reconstructed in two symmetric regions in the occipital cortex and shows a fast propagation of activity from the right toward the left, as confirmed by an intracerebral EEG [GAV 14].

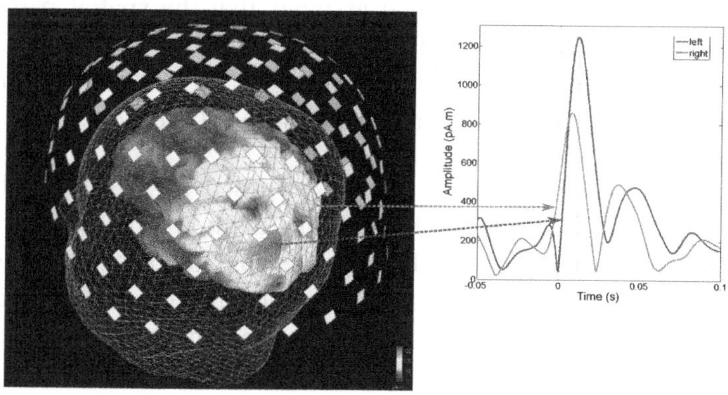

Figure 2.2. *Example of source localization through MEG on an average of eight epileptic discharges. The amplitude map is shown at the activity peak. On the right hand side, the time courses reconstructed on two homologous regions in the occipital regions show a rapid propagation from the right toward the left. For a color version of this figure, see www.iste.co.uk/clerc/interfaces1.zip*

2.2.4. *Independent component analysis*

As in fMRI, independent component analysis decomposes the signal (of dimension sensors × time) into a sum of component, each component with an associated spatial component (topography) and a time course. One of the reasons the linear model is interesting is because it makes it possible to separate activities that are mixed at the level of sensors. The difference with fMRI is that the independence constraint is defined across the temporal part of the components, and not the spatial part. This is mostly due to the fact that a large number of points is necessary in order to estimate the independence between two variables with a sufficient degree of soundness; in electrophysiology, the largest dimension is time (several thousand samples versus a few hundred sensors), whereas in fMRI space is the largest (several million voxels versus a few hundred samples).

ICA has been successfully used to remove elements like blinking, or to disentangle cognitive processes [JUN 01]. As with any method, it is important to respect mathematical assumptions. In the case of ICA, this means the separability of space and time, which assumes that the spatial part of each component does not vary in time. This assumption can be violated if there is local propagation of an activity throughout the cortex (a "wave" of activity), or if the subject's head moves with respect to the MEG sensors.

Other multivariate methods have been proposed for resolving the blind separation problem, like for example the SOBI method [BEL 97]. One of the advantages of multivariate methods is to be able to take all sensors into account at the same time, and therefore better account for noise and decrease data's dimensionality (which can help to reduce the problem of multiple comparisons).

2.2.5. *Time–frequency analysis*

One of the great advantages of electrophysiology, which comes with its temporal resolution, is the possibility of isolating oscillations that emerge at different frequencies. There is of course the classic alpha oscillations (around 10 Hz, see the first EEG observation in the 1920s), and theta oscillations (4–8 Hz), which are present in cognitive protocols. More recently, oscillations beyond 40 Hz have been shown in surface EEG, MEG and intracerebral EEG

[TAL 99]. These may be as high as 300–500 Hz for pathological oscillations in epilepsy [URR 07].

Time–frequency analysis allows characterizing such oscillations, both in terms of frequency and time duration. Several methods exist, which differ in the possibility of reconstructing data within a given window of time and frequency, as well as in their resolution in temporal and frequential dimensions. Short-term Fourier transforms are calculated with a sliding window. The greater the size of the window, the better the frequency resolution and the worse the temporal resolution will be. Another classic method is based on Morlet wavelets, which have a higher temporal resolution for higher frequencies (like in time-scale analysis). These wavelets enable a more visual representation of data, with oscillations producing characteristic time–frequency patterns [TAL 99]. On the other hand, they do not allow the reconstruction of filtered data, unlike orthogonal wavelets [LIN 14].

One of the advantages of these kinds of analyses is that, unlike simple filtering, they make it possible to study the structure of events in the time–frequency plane. In particular, it is possible to distinguish between activities that actually oscillate and transient activities that have energy at all frequencies [JMA 11].

One potential problem with these methods is that they may be sensitive to muscular activity. It has thus been shown that microsaccades that occur during the processing of a visual stimulus can produce spurious gamma activations [YUV 08].

2.2.6. *Connectivity*

Connectivity measurements seek to construct interaction graphs between different brain regions; they can be calculated on time intervals in the cortex after source reconstruction. In particular, electrophysiology measurements enable a good temporal resolution that makes it possible to measure delays between brain regions. They therefore seem to be particularly well suited for measuring oriented networks; that is, for obtaining graphs with arrows indicating the direction of information transfer from one region to another.

The simplest method is cross-correlation between two different temporal series, which measures the correlation for a series of time lapses and picks out the correlation peak. A coupling value and a time lapse are thus obtained.

Coherence is the equivalent of correlation in frequency. For each frequency, a normalized value between 0 and 1 measures the correlation level in both amplitude and phase. It is important to note that it is also possible to determine delays between the signals based on the phase slope in a given frequency band [GOT 83].

Other methods exist, like nonlinear correlation [WEN 09] or Granger causality [BRO 04]. Just like in fMRI, bivariate analyses can have limitations and it may be interesting to go on to multivariate analyses, like partial directed coherence or the directed transfer function [KU 04].

One difficulty in the application of connectivity methods comes from the existence of instantaneous correlations. At the level of sensors, these correlations come from volume conduction (one source is reflected on a large number of sensors in the same instant). At the level of sources, this is referred to as "source leakage", since one source is reconstructed with a large spatial extension. This extension comes from biophysical properties (close sources with similar orientation have very close contributions at the level of sensors) and from the resolution of the inverse problem (regularization smooths the reconstruction image). One solution for overcoming this correlation problem is to use the imaginary part of coherence [NOL 04], but this reduces the amount of information available.

2.2.7. *Statistical analysis*

In order to be able to identify the active regions in a cognitive protocol, it is necessary to calculate statistical maps in a manner similar to that used for fMRI. These maps can, for example, represent a *t*-test between two experimental conditions in each point in space for the time sample. One additional difficulty as compared to fMRI is the fact that the multiple comparison problem is even more pronounced. Indeed, tests are potentially replicated at each point in space (for example for each dipole on the cortical surface), at each point in time (typically several hundred samples) and even at each frequency.

Non-parametric methods make it possible to handle multiple comparisons (see section 2.1.3). Some have proposed using the maximum amplitude statistic among sources [PAN 03], or a "cluster" measurement that adds all statistics in t over contiguous time spans that exceed a certain threshold [MAR 07].

2.3. Simultaneous EEG-fMRI

2.3.1. *Basic principles*

As discussed in the fMRI section, the BOLD response has low temporal resolution; moreover, it is difficult to detect spontaneous events on the MRI. In order to overcome these difficulties, it can be interesting to record an EEG during an fMRI session, which may provide measurable temporal information from EEG during the MRI's data analysis [IVE 93].

Of course, this simultaneous recording is not without its technical difficulties. In particular, fMRI gradients produce large amplitude currents in EEG cables, as well as a slight movement of the head and a strong magnetic field. However, these difficulties have for the most part been resolved and it is now possible to find commercially available systems for recording simultaneous EEG-fMRI.

2.3.2. *Applications and data analysis*

One of the most important applications of EEG-fMRI is to use EEG detection of spontaneous epileptic discharges in order to analyze fMRI data. This information is useful in presurgical evaluation of patients. In this case, a linear model is constructed based on the time of observed discharges and a BOLD response model [BEN 02].

One other application on continuous data is to follow oscillation fluctuations in the frequency bands that are visible on the EEG. This makes it possible to characterize regions in the fMRI that are involved in brain rhythms, like the alpha rhythm [GOL 02].

For cognitive protocols, a strategy is based on constructing parametric regressors to bring out regions involved in different waves visible on the EEG.

In order to do this a single trial estimation of the wave amplitude (or latency) is carried out, and the amplitude of a new regressor reflects that of the parameter [BÉN 07].

In order to go further in data analysis, it is advantageous to combine the data in a joint analysis. Bayesian methods seem well suited for this task [DAU 07].

2.3.3. *Connections between EEG and fMRI*

One major question during the combination of EEG and fMRI data is the relationship between the two types of signals. In this regard, the study of neurovascular coupling is an active topic of research. Several studies have shown that the strongest connection seems to occur at the level of resting potentials and not of action potentials [LOG 01]. Moreover, this link has been observed to be more prominent in the frequency band of gamma oscillations (40–120 Hz) [NIE 05].

Numerous efforts have been carried out to represent this coupling through computational modeling [BLA 11]. Enhanced modeling techniques will eventually help to better understand the mechanisms that underlie these observations – for example negative BOLD signal fluctuations – as well as to improve data merging.

2.4. Discussion and outlook for the future

The different modalities of functional neuroimaging, EEG, MEG and fMRI, are often thought to compete with one another, but they are in fact complementary. EEG and MEG can trace back brain activity within the temporal scale of cerebral functions – that is at the millisecond level – even though they must solve an inverse problem to do so. fMRI makes it possible to directly reconstruct activity emanating from a voxel of approximately 1 cubic mm, although it is based on a relatively slow response (of about 1 s). Concerning fusion of EEG and MEG, several studies have shown that some activities are more visible in one than in the other. EEG is more sensitive to radial dipoles, whereas MEG is much less influenced by the skull and seems well suited to measure high frequencies.

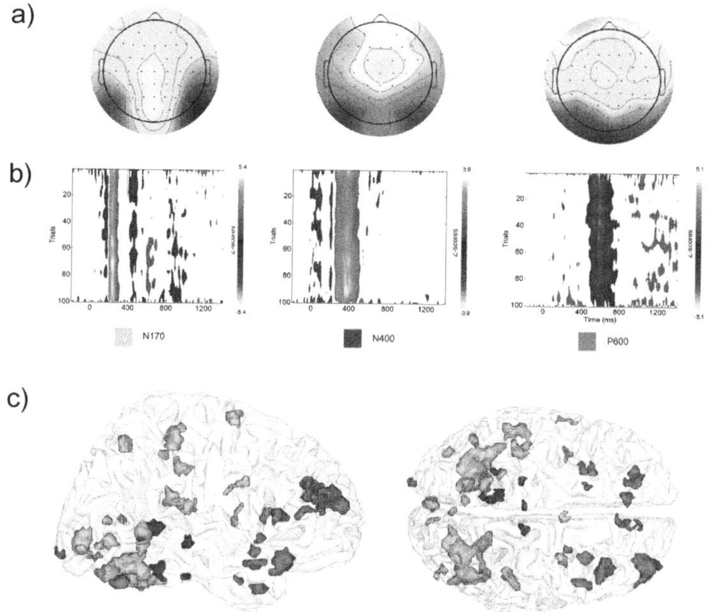

Figure 2.3. *Example of data analysis for simultaneous EEG-fMRI in a face recognition protocol (data: Jonathan Wirsich and Maxime Guye, CEMEREM-CRMBM, UMR 7339). Independent component analysis (carried out on the entire subject group) finds components corresponding to different waves observed on the EEG (N170, N400, P600). The amplitude values are then used in a parametric regression to determine the regions where the MRI signal correlates with the parameter of interest (according to [WIR 14]): a) component topography; b) visualization of tests for each component; c) results of the fMRI analysis; each color indicates the voxels for which a significant correlation with the corresponding parameter (amplitude from N170, N500 or P600) was found. For a color version of this figure, see www.iste.co.uk/clerc/interfaces1.zip*

Due to this complementarity, it seems particularly interesting to record the different imaging modalities at the same time, either EEG-MEG or EEG-fMRI. These simultaneous recordings have the great advantage of picking up the same brain activity from different angles, but they also present new technical challenges for the analysis of this complex data.

Among several avenues for future research, we can highlight the importance of confirming non-invasive measures with invasive measures

[DUB 14], and improving data interpretation with biophysical and computational modeling [VOG 12].

2.5. Bibliography

[BAI 01] BAILLET S., MOSHER J.C., LEAHY R.M., "Electromagnetic brain mapping", *IEEE Signal Processing Magazine*, vol. 18, pp. 14–30, 2001.

[BEN 02] BÉNAR C.G., GROSS D.W., WANG Y. *et al.*, "The BOLD response to interictal epileptiform discharges", *Neuroimage*, vol. 17, pp. 1182–1192, 2002.

[BEL 97] BELOUCHRANI A.A., MERAIM K., CARDOSO J.F. *et al.*, "A blind source separation technique using second order statistics", *IEEE Transactions on Signal Processing*, vol. 45, no. 434, p. 444, 1997.

[BET 11] BETTUS G., RANJEVA J.P., WENDLING F. *et al.*, "Interictal functional connectivity of human epileptic networks assessed by intracerebral EEG and BOLD signal fluctuations", *PLoS One*, vol. 6, p. e20071, 2011.

[BLA 11] BLANCHARD S., PAPADOPOULO T., BÉNAR C.G. *et al.*, "Relationship between flow and metabolism in BOLD signals: insights from biophysical models", *Brain Topography*, vol. 24, pp. 40–53, 2011.

[BRO 04] BROVELLI A., DING M., LEDBERG A. *et al.*, "Beta oscillations in a large-scale sensorimotor cortical network: directional influences revealed by Granger causality", *Proceedings of the National Academy of Sciences*, vol. 101, pp. 9849–9854, 2004.

[BUR 98] BUROCK M.A., BUCKNER R.L., WOLDORFF M.G. *et al.*, "Randomized event-related experimental designs allow for extremely rapid presentation rates using functional MRI", *Neuroreport*, vol. 9, pp. 3735–3739, 1998.

[BÉN 07] BÉNAR C.G., SCHÖN D., GRIMAULT S. *et al.*, "Single-trial analysis of oddball event-related potentials in simultaneous EEG-fMRI", *Human Brain Mapping*, vol. 28, pp. 602–613, 2007.

[COL 96] COLES M., RUGG, M., "Event-related brain potentials: an introduction", in *Electrophysiology of Mind*, Oxford University Press, 1996.

[COM 94] COMON P., "Independent component analysis: a new concept?", *Signal Processing*, vol. 36, pp. 287–314, 1994.

[DAM 06] DAMOISEAUX J.S., ROMBOUTS S.A., BARKHOF F. *et al.*, "Consistent resting-state networks across healthy subjects", *Proceedings of the National Academy of Sciences*, vol. 103, pp. 13848–13853, 2006.

[DAU 07] DAUNIZEAU J., GROVA C., MARRELEC G., *et al.*, "Symmetrical event-related EEG/fMRI information fusion in a variational Bayesian framework", *Neuroimage*, vol. 36, pp. 69–87, 2007.

[DAV 13] DAVEY C.E., GRAYDEN D.B., GAVRILESCU M. *et al.*, "The equivalence of linear Gaussian connectivity techniques", *Human Brain Mapping*, vol. 34, 1999–2014, 2013.

[DUB 14] DUBARRY A.S., BADIER J.M., TRÉBUCHON-DA FONSECA A. *et al.*, "Simultaneous recording of MEG, EEG and intracerebral EEG during visual stimulation: from feasibility to single-trial analysis", *Neuroimage*, vol. 1, no. 99, pp. 548–58, 2014.

[FRI 94] FRISTON K.J., WORSLEY K.J., FRACKOWIAK R.S.J. *et al.*, "Assessing the significance of focal activations using their spatial extent", *Human Brain Mapping*, vol.1, pp. 214–220, 1994.

[GAV 14] GAVARET M., BADIER J.M., BARTOLOMEI F. *et al.*, "MEG and EEG sensitivity in a case of medial occipital epilepsy", *Brain Topography*, vol. 27, pp. 192–196, 2014.

[GEN 02] GENOVESE C.R., LAZAR N.A., NICHOLS T., "Thresholding of statistical maps in functional neuroimaging using the false discovery rate", *Neuroimage*, vol. 15, pp. 870–878, 2002.

[GOL 02] GOLDMAN R.I., STERN J.M., ENGEL J., COHEN M.S., "Simultaneous EEG and fMRI of the alpha rhythm", *Neuroreport.*, vol. 13, no. 18, pp. 2487–2492, 2002.

[GOT 83] GOTMAN J., "Measurement of small time differences between EEG channels: method and application to epileptic seizure propagation", *Electroencephalography and Clinical Neurophysiology*, vol. 56, pp. 501–514, 1983.

[HOG 99] HOGE R.D., ATKINSON J., GILL B. *et al.*, "Investigation of BOLD signal dependence on cerebral blood flow and oxygen consumption: the deoxyhemoglobin dilution model", *Magnetic Resonance in Medicine*, vol. 42, pp. 849–863, 1999.

[HOL 06] HOLM A., RANTA-AHO P.O., SALLINEN M. *et al.*, "Relationship of P300 single-trial responses with reaction time and preceding stimulus sequence", *International Journal of Psychophysiology*, vol. 61, pp. 244–252, 2006.

[HOR 96] HORNAK J., The Basics of MRI, available at https://www.cis.rit.edu/htbooks/mri/, 1996.

[HÄM 93] HÄMÄLÄINEN M., HARI R., ILMONIEMI R.J. *et al* "Magnetoencephalography – theory, instrumentation, and applications to noninvasive studies of the working human brain", *Reviews of Modern Physics*, vol. 65, pp. 414–497, 1993.

[HÄM 94] HÄMÄLÄINEN M.S., ILMONIEMI R.J., "Interpreting magnetic fields of the brain: minimum norm estimates", *Medical and Biological Engineering and Computing*, vol. 32, pp. 35–42, 1994.

[IVE 93] IVES J.R., WARACH S., SCHMITT F. *et al.*, "Monitoring the patient's EEG during echo planar MRI", *Electroencephalography and Clinical Neurophysiology*, vol. 87, pp. 417–420, 1993.

[JMA 11] JMAIL N., GAVARET M., WENDLING F. *et al.*, "A comparison of methods for separation of transient and oscillatory signals in EEG", *Journal of Neuroscience Methods*, vol. 199, pp. 273–289, 2011.

[JUN 01] JUNG T.P., MAKEIG S., WESTERFIELD M. *et al.*, "Analysis and visualization of single-trial event-related potentials", *Human Brain Mapping*, vol. 14, pp. 166–185, 2001.

[KUŚ 04] KU R., KAMISKI M., BLINOWSKA K.J. "Determination of EEG activity propagation: pair-wise versus multichannel estimate", *IEEE Transactions on Biomedical Engineering*, vol. 51, pp. 1501–510, 2004.

[LIN 14] LINA J.M., CHOWDHURY R., LEMAY E. *et al.*, "Wavelet-based localization of oscillatory sources from magnetoencephalography data", *IEEE Transactions on Biomedical Engineering*, vol. 61, pp. 2350–364, 2014.

[LIN 01] LOGOTHETIS N.K., PAULS J., AUGATH M. *et al.*, "Neurophysiological investigation of the basis of the fMRI signal", *Nature*, vol. 412, pp. 150–157, 2001.

[LOP 05] LOPES DA SILVA F.H., VAN ROTTERDAM A., "Biophysical aspects of EEG and magnetoencephalographic generation", in NIEDERMEYER E., LOPES DA SILVA F.H., (eds), *Electroencephalography: Basic Principles, Clinical Applications and Related Fields*, Lippincott, Williams & Wilkins, New York, 2005.

[MAR 07] MARIS E., OOSTENVELD R., "Nonparametric statistical testing of EEG- and MEG-data", *Journal of Neuroscience Methods*, vol. 164, pp. 177–190, 2007.

[MAR 06] MARRELEC G., KRAINIK A., DUFFAU H. *et al.*, "Partial correlation for functional brain interactivity investigation in functional MRI", *Neuroimage*, vol. 32, pp. 228–237, 2006.

[MCK 98] MCKEOWN M.J., MAKEIG S., BROWN G.G. *et al.*, "Analysis of fMRI data by blind separation into independent spatial components", *Human Brain Mapping*, vol. 6, pp. 160–188, 1998.

[NIC 03] NICHOLS T., HAYASAKA S., "Controlling the familywise error rate in functional neuroimaging: a comparative review", *Statistical Methods in Medical Research*, vol. 12, pp. 419–446, 2003.

[NIC 02] NICHOLS T., HOLMES A.P., "Nonparametric permutation tests for functional neuroimaging: a primer with examples", *Human Brain Mapping*, vol. 15, pp. 1–25, 2002.

[NIE 05] NIESSING J., EBISCH B., SCHMIDT K.E. *et al.*, "Hemodynamic signals correlate tightly with synchronized gamma oscillations", *Science*, vol. 309, pp. 948–951, 2005.

[NOL 04] NOLTE G., BAI O., WHEATON L. *et al.*, "Identifying true brain interaction from EEG data using the imaginary part of coherency", *Clinical Neurophysiology*, vol. 115, pp. 2292–307, 2004.

[NUN 05] NUNEZ, P., SRINIVASAN R., *Electric Fields of the Brain: The Neurophysics of EEG*, Oxford University Press, Oxford, 2005.

[OGA 90] OGAWA S., LEE T.M., NAYAK A.S. *et al.*, "Oxygenation-sensitive contrast in magnetic resonance image of rodent brain at high magnetic fields," *Magnetic Resonance in Medicine*, vol. 14, pp. 68–78, 1990.

[PAN 03] PANTAZIS D., NICHOLS T.E., BAILLET S. *et al.* "Spatiotemporal localization of significant activation in MEG using permutation tests", *Information Processing in Medical Imaging*, vol. 18, pp. 512–523, 2003.

[PEN 04] PENNY W.D., STEPHAN K.E., MECHELLI A. *et al.*, "Modelling functional integration: a comparison of structural equation and dynamic causal models", *Neuroimage*, vol. 23, no. Suppl 1, pp. S264–S274, 2004.

[PER 07] PERLBARG V., BELLEC P., ANTON J.L. *et al.* "CORSICA: correction of structured noise in fMRI by automatic identification of ICA components", *Magnetic Resonance Imaginging*, vol. 25, pp. 35–46, 2007.

[RAI 07] RAICHLE M.E., SNYDER A.Z., "A default mode of brain function: a brief history of an evolving idea", *Neuroimage*, vol. 37, pp. 1083–1090; discussion 1097–1099, 2007.

[ROE 05] ROEBROECK A., FORMISANO E., GOEBEL R., "Mapping directed influence over the brain using Granger causality and fMRI", *Neuroimage*, vol. 25, pp. 230–242, 2005.

[SCH 85] SCHERG M., VON CRAMON D., "Two bilateral sources of the late AEP as identified by a spatio-temporal dipole model", *Electroencephalography and Clinical Neurophysiology*, vol. 62, pp. 32–44, 1985.

[SMI 11] SMITH S.M., MILLER K.L., SALIMI-KHORSHIDI G. *et al.*, "Network modeling methods for FMRI", *Neuroimage*, vol. 54, pp. 875–891, 2011.

[TOL 99] TALLON-BAUDRY C., BERTRAND O., "Oscillatory gamma activity in humans and its role in object representation", *Trends Cognitive Science*, vol. 3, pp. 151–162, 1999.

[URR 07] URRESTARAZU E., CHANDER R., DUBEAU F. *et al.*, "Interictal high-frequency oscillations (100-500 Hz) in the intracerebral EEG of epileptic patients", *Brain*, vol. 130, pp. 2354–2366, 2007.

[VAN 88] VAN VEEN B., BUCKLEY K., "Beamforming: a versatile approach to spatial filtering", *IEEE ASSP Magazine*, vol. 5, pp. 4–24, 1988.

[VOG 12] VOGES N., BLANCHARD S., WENDLING F. *et al.*, "Modeling of the neurovascular coupling in epileptic discharges", *Brain Topography*, vol. 25, no. 2, pp. 136–156, 2012.

[WEN 09] WENDLING F., ANSARI-ASL K., BARTOLOMEI F. *et al.*, "From EEG signals to brain connectivity: a model-based evaluation of interdependence measures", *Journal of Neuroscience Methods*, vol. 183, pp. 9–18, 2009.

[WIR 14] WIRSICH J., BÉNAR C., RANJEVA J.P. *et al.*. "Single-trial EEG-informed fMRI reveals spatial dependency of BOLD signal on early and late IC-ERP amplitudes during face recognition", *Neuroimage*, vol. 100, pp. 325–336, 2009.

[WOR 02] WORSLEY K.J., LIAO C.H., ASTON J. *et al.*, "A general statistical analysis for fMRI data", *Neuroimage*, vol. 15, pp. 1–15, 2002.

[WOR] WORSLEY K.J., MARRETT S., NEELIN P. *et al.*, "A unified statistical approach for determining significant signals in images of cerebral activation", *Human Brain Mapping*, vol. 4, pp. 58–73, 1996.

[YUV 08] YUVAL-GREENBERG S., TOMER O., KEREN A.S. *et al.*, "Transient induced gamma-band response in EEG as a manifestation of miniature saccades", *Neuron*, vol. 58, pp. 429–441, 2008.

3

Cerebral Electrogenesis

Hans Berger was the first to use electroencephalography to record electric brain activity in humans at the end of the 1920s [BER 29]. However, the idea that these signals do indeed originate in the brain was not immediately accepted, and only after Lord Adrian, the eminent British physiologist from Cambridge, reproduced Berger's results at the beginning of the 1930s, did the scientific community become interested in the technology [ADR 34]. Since then, medical and scientific applications of electroencephalography have proven to be of considerable importance.

However, although it is now well established that electroencephalogram (EEG) measurements record brain activity, it is important to note that they only reveal a small portion of brain activity. Understanding what type of brain activity can or cannot be revealed by EEG is fundamental (1) in order to accurately interpret EEG recordings and (2) in order to know what information they can (or cannot) provide.

3.1. Electrical neuronal activity detected in EEG

For the most part, electrical brain activity is produced by neurons, so EEG essentially records neuronal activity. The following paragraphs describe this activity in detail.

Chapter written by Franck VIDAL.

3.1.1. *Action and postsynaptic potentials*

Electrical neuronal activity includes action potentials (APs) and postsynaptic potentials (PSPs).

APs are produced at the initial segment of the axon, from which they propagate in an active (regenerative) manner toward the peripheral end, without being attenuated, at a speed (ranging from a few meters to several dozen meters per second) that depends on the axon's myelination and diameter. A typical action potential lasts between 1 and 3 ms for an amplitude of about 100 mV. PSPs are produced at the synapses, in the dendrites and the neuron cell bodies[1]. For the most part, they propagate instantaneously, and in a passive (non-regenerative) manner[2] and for that reason, they attenuate with distance. Unlike APs, PSP amplitudes are not fixed, but rather depend on the intensity of synaptic activity. However, their amplitude remains far weaker than that of APs around 10 mV. On the other hand, PSPs last far longer than APs in the order of 20 ms [KAN 00]. At first sight, we might think that APs contribute most to EEG, since they are far stronger than PSPs. However, we will see that their short duration makes their contribution rather weak. We will see, on the other hand, that the relatively long duration of PSPs makes them prime contributors to EEG activities, despite their low amplitude.

EEG records brain activity at a distance through the bone and the scalp. Metaphorically, it is similar to hearing a conversation between people located far away, on the other side of a wall. A single person speaking loudly would not be heard. Several people speaking at a normal volume while taking turns would not be heard either. Instead, several people speaking at a normal volume *at the same time* would be audible. In the same manner, it is necessary for several neurons to be active at the same time in order for the sum of their elementary activities to reach an intensity sufficient to be measured at the surface of the scalp. In other words, EEG is only sensitive to the activity of neurons when enough of them are active at the same time.

---

1 There are also synapses between a dendrite and an axon.

2 There are also cases of regenerative transmission outside of the axonal compartment, but these are the exception.

When an afferent volley reaches a certain structure, it activates its target neurons synchronously. However, taking into account the inherent variability of any biological phenomenon, this synchrony is not perfect and the resulting activity is only *quasi*-synchronous. Given this variability, APs' short lifespan makes their summation unlikely[3]. On the other hand, even when they are set off in an imperfectly synchronous manner, PSPs last long enough to effectively be added-up, even if a part of their activity is lost in the variability of neuronal activity.

It is therefore PSPs that produce the most important contribution to EEG activity, and we will study them in more detail.

3.1.2. *Resting potential, electrochemical gradient and PSPs*

Active transport of some ions through the plasma membrane, on the one hand, and differences in membrane permeabilities to different ions, on the other hand, produce differences in concentrations among intra- and extracellular compartments, as well as polarization in the membrane, which becomes positively charged on the outside and negatively charged on the inside [KAN 00]. The difference in transmembrane potential due to this asymmetric distribution of charges constitutes a neuron's "resting potential".

Differences in concentration between intra- and extracellular compartments create a chemical gradient that drives ions to move from the most concentrated compartment to the least concentrated. Moreover, the resting potential pushes them toward (or keeps them in) the opposite charge compartment. The combination of these chemical and electric gradients, which is known as an "electrochemical gradient", determines the direction (and the strength) of each ion's spontaneous tendency to move toward one compartment or another. As an example, the potassium ion, since it is positively charged, is retained by the electrical gradient in the intracellular milieu; but that same potassium ion, since it is more concentrated in the intracellular milieu, is pushed by its chemical gradient to the extracellular milieu. Potassium's chemical gradient being stronger than its electrical

3 We will see later that it is also necessary to consider the "quadrupole" nature of the field produced by classic axonal APs, which attenuates strongly with distance from the recording position, thus further reducing their contribution to EEG.

gradient, the resulting electrochemical gradient pushes it toward the extracellular compartment.

The release of a neurotransmitter by the presynaptic neuron directly or indirectly produces the opening of ion channels on the postsynaptic neuron. Opening these channels makes it possible for certain ions to pass from one compartment to another in the direction determined by their electrochemical gradient [KAN 00]. For example, opening potassium channels makes it possible for that ion to follow its electrochemical gradient toward the outside. The outflow of positive charge that results from this process increases polarization of the membrane and thereby contributes to making the neuron more difficult to excite. For this reason, PSPs resulting from opening the permeable channels are called "inhibitory". But they are not the only ones, and any channel whose opening "overpolarizes" the membrane generates, when it is opened, inhibitory postsynaptic potentials (IPSP). On the other hand, opening channels that are mainly permeable to sodium[4] makes it possible for that ion to follow the strong electrical and chemical gradients that push it toward the intracellular compartment. The resulting net inflow of positive charges reduces the difference in potentials between the interior and the exterior: it depolarizes the membrane and thereby contributes to making the neuron more easily excitable. For this reason, PSPs resulting from the opening of channels that are mainly permeable to sodium are called "excitatory". But they are not the only ones, and all channels whose opening depolarizes the membrane generate excitatory postsynaptic potential (EPSP) when they are opened.

3.1.3. *From PSPs to EEG*

Let us consider the case of EPSPs generated by opening the channels that are mainly permeable to sodium. Opening the channels enables an influx of sodium ions and that movement of positive charges constitutes an incoming current. That so-called "primary" incoming current is matched by an equivalent outgoing current, through the neuronal membrane (since the cell membrane's resistance is not infinite), distributed along the dendrites and the cell body while attenuating with distance. Incoming and outgoing currents,

4 In the following, we will neglect their permeability to potassium of those channels, since when they are open, the most important effect pertains to sodium.

which are known as "imposed", produce an intracellular diffusion current, as well as an electrical field in the surrounding extracellular milieu [PER 07]. Since the extracellular milieu is conductive, the electrical field generated by the imposed currents also induces currents, which are passively conducted in the extracellular milieu, and called "secondary" currents. They circulate throughout the volume of the head in such a way that the current lines close, thereby respecting the charge conservation principle (see Figure 3.1). If these secondary currents reach the scalp and are sufficiently strong, they produce potential variations that are measurable by the EEG.

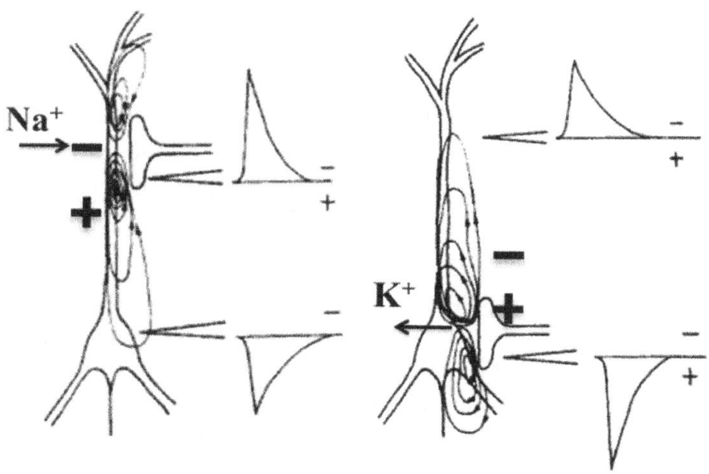

Figure 3.1. *Left: Currents created by an inflow of sodium and potential, as recorded by nearby distal electrodes. Right: Currents created and potentials recorded during a deep outflow of potassium*

The opening of ion channels is responsible for EPSPs and IPSPs[5], thereby generating the primary currents that give way to secondary currents responsible for the activities recorded by EEG by enabling ions to move in the direction determined by their electrochemical gradient.

While opening the (dendritic or somatic) ionic channels is indeed the primary cause of measurable differences in potentials recorded in EEG, we

5 It is necessary to also consider, in some cases, the contribution of specific channels depending on an intracellular ligand, which we will explore later (section 2.1.3).

should not imagine that any opening of ion channels, even when massive and physiologically pertinent, gives way to measurable electrical activity. Indeed, opening certain ionic channels, even when it produces a strong inhibitory effect, does not generate (or only weakly so) a PSP. This is the case when channels permeable to the chloride ion are opened. For many neurons, the chloride ion's electrochemical gradient is null or close to zero. For this reason, at or near the resting potential, massive opening of the channels permeable to chloride does not produce (or barely does so) a net flow of chloride through the membrane and therefore little or no variations in PSP are produced. On the other hand, this massive opening considerably increases the membrane's electrical conductance and, through shunting, it impedes or strongly attenuates the depolarizing effect induced by opening other channels (channels permeable to sodium, for example)[6] [KAN 00] (the majority of these channels are also permeable to the bicarbonate ion, but since their permeability to chloride is dominant, we have ignored the effect of bicarbonates – even though it is not without its functional consequences). A substantial part of the very real inhibition exerted by the chloride channels opening is therefore electrically silent; the EEG is blind to it. (Note, however, that if the neuron is depolarized or hyperpolarized, a significant electrochemical gradient for the chloride ion appears at that moment; if at the same time a synaptic activity provokes the opening of the chloride channels, it will enable a net transmembrane flow of chloride ions responsible for significant primary and secondary currents and whose direction will depend on the sign of the electrochemical gradient, determined at that moment, for the chloride ion).

Finally, there are electrical synapses which involve gap junctions: currents circulate from one neuron to the other through the junctions formed by the butt-jointed association of two pores (one per neuron). The junction of two pores forms a little "tunnel" (each pore forms half of the tunnel) that crosses each neuron's membrane as well as the extracellular milieu that separates them, thereby making the intracellular milieus of the two neurons communicate. Since the currents remain confined to the intracellular compartment, they do not generate significant extracellular currents

6 In the case of a sodium inflow, prior opening of channels permeable to Cl^- allows those ions to passively follow Na^+ toward the interior. In other words, during an incoming current carried by intake of Na^+, the previous opening of the Cl^- channels provides an outward leakage of current (shunt) carried by the inflow of Cl^- inflow.

[BUZ 12]. As a result, the electrical synapses do not contribute significantly to the EEG. On the other hand, those synapses often participate in the electrical coupling of neuronal ensembles; in this sense, they increase synchrony between the activities of connected neurons through those synapses, which facilitates summation of their activities and contributes to increase the amplitude of signals measured at a distance.

3.2. Dipolar and quadrupole fields

3.2.1. *Field created by an ion current due to the opening of ion channels*

We have indicated above that opening of ion channels, when they give way to ion currents, is accompanied by currents of the same intensity that cross the membrane in the opposite direction. The currents generated by opening ion channels (almost) always go through a single point (or points), whereas those that cross the membrane in the opposite direction are distributed throughout the neural membrane and attenuate with distance (see Figure 3.1).

Let us go back to the example of sodium inflow: an extracellular electrode will "see" current disappearing from the extracellular milieu wherever there are open channels, referred to as a "current sink". At the same time, the electrode will see current appear throughout the rest of the neural membrane, which is referred to as a "current source". An extracellular electrode thus detects the association of a current sink and a current source of the same intensity. Having equal intensity and opposite signs, if both the sink and the source occurred at individual points, they would produce a current dipole ; but since the source is distributed throughout the membrane, this sink–source configuration does not quite produce a current dipole.

In fact, the potential created at a point by such a sink–source configuration is given by an infinite sum, whose first term corresponds to the potential created by a dipole, the second to the potential created by a quadrupole, the third to the potential created by an octupole, etc[7]. [PER 07]. We should note that for a given entering current, the greater the distance between the sink and

7 In the case that interests us, according to the principle of charge conservation, charge is neither created nor destroyed in the brain; the monopolar term is therefore equal to zero and does not appear in the expression of potential created by the opening of ion channels.

the barycenter of the distributed sources, the greater the potential captured by the dipolar term. Moreover, the dipolar term decreases proportional to the square of the distance to the point where measurement is taken. The quadrupole term decreases proportional to the cube of this same distance. The octupolar term decreases proportional to this distance raised to the fourth power. Therefore, if the distance between the sink and the barycenter of the sources is large enough (on a neural scale), and if the recording point is far enough, as is the case for EEG (but not necessarily when recording local field potentials), the dipolar term dominates all the others, which become negligible for practical purposes. On the other hand, if the distance between the sink and the source's barycenter is negligible or zero, the quadrupole term will dominate.

3.2.1.1. *Field created by an inflow of ions during a synapse (PSP)*

The geometry of dendrites and cell bodies is not simple, but if a neuron presents an asymmetry in its dendrites it is often possible to account for synaptic activities using a relatively simple model; this is the case for the cerebral cortex's pyramidal neurons, which we will use as an example. Let us first specify that these neurons are roughly shaped like a triangle whose base is oriented toward the deep layers of the cortex and the top toward its outer layers. They have a small basal dendritic arborization, but most importantly, they have one very long dendrite called the "apical dendrite". It stems from the apex of the triangle toward the outer layers of the cortex, being somewhat ramified in its distal end and perpendicular to the cortex's surface. A sodium inflow occurring at the apex (or the base) of the apical dendrite of a pyramidal neuron produces a point sink and a source distributed throughout the rest of the dendrite and the cell body (Figure 3.1). The current will emerge more from one side than the other (because there is more surface available for current leakage on one side than on the other). The distributed source's barycenter will therefore not coincide with the sink and, all other things being equal, even less so when we consider that the inflow of sodium will occur near one of the dendrite's ends. The dipolar term will therefore dominate the expression of potential created by the sodium inflow, and approximating this kind of generator with a current dipole is reasonable. This is why the dipolar model is often used to model the sources of EEG activity, which are essentially due to synaptic activity, that is to say to PSP. One should bear in mind that it is a model that relies on the kind of approximations we have just described.

If the dendrite's shape is similar to a straight line, which is often the case for apical dendrites in the pyramidal neurons, then the formula for the potential, $V(P)$, created by such a dipole at a point P (see section 3.2), is rather simple:

$$V(P) = \frac{1}{4\pi\sigma}\frac{M\cos(\theta)}{r^2} \qquad [3.1]$$

where σ is the medium's conductivity, M is the module of the dipole's moment, r is the distance from point P to the middle of the segment connecting the sink and the barycenter of the source and θ is the angle formed by the dipole and the segment connecting its midpoint to the point P (Figure 3.2) [PER 07]. This expression relies on an additional approximation made possible by representing the apical dendrite as a straight line because it is possible to assume that the distribution of the potential generated by an inflow of sodium is symmetric by rotation [PER 07]. For more complex dendritic geometries, where this assumption would be invalid, the formula would be less simple.

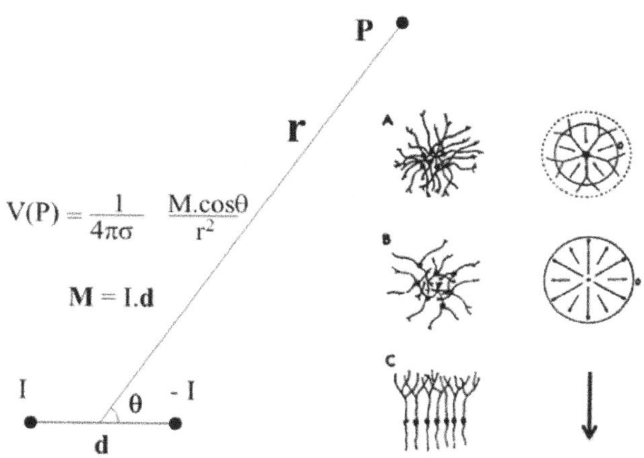

Figure 3.2. *Left: Potential created by a current dipole. Right: a) Null equivalent dipoles resulting from stellate neuron activation b) or from neurons without a specific orientation; c) non-zero equivalent dipole resulting from the activation of parallel-dendrite neurons (adapted from [LOP 82])*

3.2.1.2. *Field created by an ion inflow at the axon (AP)*

Let us stick to the example of sodium inflow, but this time, during the propagation of the AP in the axon. The inflow of sodium through the voltage-gated ion channels produces an outflow of current on both sides of the sink. Since (1) it is possible to locally represent an axon as a straight line, and (2) the same amount of current outflows on the left and the right hand sides of the sink, the distributed source's barycenter is located at the exact same point as the sink. The distance between the source's barycenter and the sink being equal to zero, the dipolar term is also zero. The dominant term in the expression of the potential generated by an AP is therefore the following term in the series, namely the quadrupole term. We have seen that it decreases proportional to the cube of the distance between the generator and the recording point. This very sharp decrease with distance is one of the reasons why APs contribute little to EEG, as compared to PSP. Another reason, as we have seen above, has to do with the fact that typical APs do not last long enough for their activity to be added together effectively.

A third reason can be added to the two previous ones: for the most part, axons have a small diameter (as compared to that of dendrites or the soma); their membrane's external surface is therefore relatively small and there is therefore relatively little surface available for transmembrane currents to pass. The transmembrane resistance of a portion of a given axon is thus relatively high. For this reason, despite high AP values (of about 100 mV), they do not produce intense currents (for a more detailed explanation see Chapter 8 in this volume, which studies extracellular recordings).

Let us also note that in the case of local field potentials, the recording electrode (which is intracerebral) can be very close to the AP or to a set of APs. In that case, since the electrode is close to the sink, potential attenuation at the recording point, even when it is proportional to the cube of the distance to the sink, can remain relatively low. Moreover, since intracerebral currents are not attenuated by the skull's bones, there is no need for as large a summation as for EEG in order for the electrode to record measurable activity. Therefore, when an afferent volley activates a large number of neurons, the resulting activity may be large enough to be measurable as local field potentials, even though only a small portion of the activity generated by the APs is actually summated; more so, if the electrode is near the generators.

3.2.1.3. *Field created by other neuronal activities*

There may also be some neurons for which particular APs do not depend on channels permeable to sodium, but rather on other ion channels, like, for example calcium. These channels permeable to calcium are essentially dendritic and, such as traditional APs, the APs they generate have large amplitudes. But unlike traditional APs, they are long lasting: ten to several tens of milliseconds [HEL 99]. They therefore accumulate easily. Although it is clear that they contribute to local field potentials, it is more difficult to understand their contribution to EEG, since there is not currently enough data about their activity *in vivo* [BUZ 12]. For the moment, we can only conclude that the possibility of calcium APs contributing to EEG is not ruled out.

In some neurons, when the opening of certain ion channels that depend on an intracellular ligand generates a net flow of ions, it generates primary currents that induce secondary extracellular currents. For example, calcium APs do not only depolarize the membrane, but they also increase intracellular calcium content. This increase in calcium concentration has an effect on potassium channels sensitive to intracellular calcium, which are located on the soma. Opening these channels allows an outflow of calcium that produces a primary hyperpolarizing current source [HOT 80] located at the cell body. The amplitude and the duration of these non-synaptic currents are of the same order of magnitude as those of PSP. Therefore, since these calcium-dependent potassium currents are located at the cell body, the position of the source most probably differs from the associated distributed sink's barycenter. The generated field will thus have a dipolar form and it may contribute to the EEG, in the same way as it does to PSP, provided that neurons activated in this manner are sufficiently many and sufficiently synchronous.

Some types of voltage-dependent channels, other than the sodium and potassium channels responsible for the APs, can confer electric resonance properties to some neurons [LLI 88]; that is to say that those neurons respond more effectively if the afferent volleys develop in a specific frequency range that represents their electric resonance frequency. Moreover, for sufficiently high depolarization levels, the activation of such channels can bring about self-sustained membrane potential oscillations in those same neurons. For example, it is their presence that allows thalamocortical neurons to generate periodic bursts of APs in relation to sleep spindles [STE 88]; such channels are also essential in the genesis of cortical oscillations of the gamma band

[LLI 07] and of slower EEG frequencies (0.5–1 Hz) during slow-wave sleep [LLI 06]. The contribution of these non-synaptic oscillatory activities to local field potentials is likely, but even though they are necessary for producing certain cortical rhythms as we noted above, their direct contribution to EEG has not yet been completely determined. It will depend on whether the currents generated by the channels on the dendritic and somatic membranes are more or less distributed, on the number of neurons involved and on the phase synchrony of those oscillations.

3.2.2. *Factors determining the value of the potential created by an ion current*

When positioned at a given point, in a given direction, a dipole is completely characterized by its moment. That moment is represented by the vector given by the segment connecting the sink and the source, oriented in the direction going from the sink to the source, and whose module is given by the product of the current intensity and the distance separating the sink from the source. The intensity of the current depends only on the channels' conductance, their number and the electrochemical gradient for the ions carrying the current. The distance depends on the synapse's position (or on the non-synaptic sources/sinks, such as the calcium-dependent potassium currents as mentioned above) and on several factors related to the neuron's shape.

Concerning the neuron's shape, we can mention the following non-exhaustive factors:

1) the dendrite's length will determine how far from its entry point the current can outflow;

2) the number, nature and branching points of that dendrite will determine the possible leakage areas for the exiting current;

3) the dendrite's diameter is also an important factor. Indeed, since the current follows the lines of least resistance, it tends to follow the longitudinal direction (parallel to the dendrite's axis) since the lipid membrane is far more resistive than the intracellular milieu. Therefore, the greater the diameter of the dendrite, the lower the longitudinal resistance; the more the current can follow that direction, the more it can propagate (and exit or enter) farther from its point of entry.

Concerning the synapse's position, let us recall that if it is located on the distal end of a dendrite, the current leaks far more in the direction of the cell body, since there is little surface available on the distal side to allow a current leak; if the synapse is located on the proximal side, near the cell body, the current leaks far more in distal direction because there is, on that side, more surface available for current leakage. In both cases, if we consider the example of a sodium inflow, the distributed source's barycenter is an appreciable distance away from the sink and the field thereby created is very much like that created by a dipole current. Let us now consider an otherwise similar synapse located at the middle of the dendrite: the exiting current leaks as much on one side of the sink as on the other, and the barycenter of the distributed source is at the same place as the sink. Their distance being equal to zero, the dipolar term is equal to zero and the dipolar model is no longer valid. As in the case of an AP, the first non-zero term of the series, which describes the potential generated by such a synapse, is the quadrupole term, which becomes the dominant term compared to the following ones, which decrease even more sharply with distance. Just like that of APs, the activity generated by this kind of synapse contributes very little to EEG. All other things being equal, the more the synapses are positioned asymmetrically on the dendrite, far from the position (often close to the middle) at which the source's barycenter is at the same place as the sink, the greater the intensity of the field they generate.

3.3. The importance of geometry

3.3.1. *Spatial summation, closed fields and open fields*

We have seen above that in order to generate an electrical activity measurable with an EEG, the activities of several neurons must occur at the same time, which in turn allows them to summuate. This is a necessary condition, but it is not sufficient. We will now see that a spatial summation is also involved, and its effects at a distance depend on the geometry of the set of neurons activated in synchrony.

Let us assume that the simplest dipolar model, whose formula we provided above, applies. Each synaptic activity generates a current dipole that is characterized by its moment. The moment being a vector quantity, it is possible to account for the sum of several elementary synaptic activities with a so-called "equivalent" dipole whose moment corresponds to the sum of moments for each elementary dipole.

Let us now consider a stellate neuron[8] receiving a synchronous afferent volley that activates several synapses on its several dendrites. A dipole current then appears on each dendrite. Let us assume that those dipoles have roughly the same moment and that the afferent volley is distributed throughout the set of dendrites in a roughly homogeneous manner. Since the dendrites of a stellate neuron radiate in may different directions away from the cell body, the equivalent dipole moment resulting from the sum of the elementary dipoles is equal to zero or almost zero [LOR 47] (Figure 3.2 on the right (a)). We will obtain the same result with one or several synapses arranged along the cell body, since the leak current will be distributed in a regular manner along the entire the dendritic tree. Therefore, due the geometry of such neurons, an afferent volley does not produce a perceptible field in the conditions where the dipolar model could be applied, that is, for observations carried out far from the neuron (as is always the case for surface EEGs), even if it generates strong local currents. These kinds of neurons are said to produce "closed fields". Even if a population of several stellate neurons is activated synchronously as we have indicated, the fields they generate do not produce enough of a current to be detected at a distance, and for that reason they do not have a detectable influence on EEG either [LOR 47].

Let us now consider a structure containing neurons with asymmetries in their dendritic tree. If this structure receives an afferent volley, it can generate a resulting non-zero elementary dipole current at each neuron. However, if those neurons are not oriented in a predominant direction, that is to say that each neuron's dendritic tree can point to any direction, the moments of dipoles generated by the activation of that structure's neurons will also be oriented in all directions. If we assume that the moments of elementary dipoles generated by the activation of that structure's neurons are more or less the same, the equivalent dipole – which is the sum of the elementary dipoles describing the activity of all of the structure's activated neurons – will be close to zero. A cytoarchitectonic organization of this kind will also produce closed fields [LOR 47] (Figure 3.2 on the right (b)). As a result, the majority

8 These neurons have several dendrites that extend in many directions away from the cell body, accounting for their name.

of the cortical interneurons, since they tend not to be arranged in a given direction, barely contribute to EEG, even if they are strongly activated and generate intense local currents.

In order for the equivalent dipole corresponding to the synchronous activation of a large number of neurons to generate an intense field at a distance, it is necessary for those neurons to be arranged in a cytoarchitectonic organization that favors spatial summation of their activities [LOR 47] (Figure 3.2 on the right (c)). This is the case for pyramidal neurons in the cerebral cortex. In a given cortical area, the apical dendrites from those neurons are almost all parallel. If one afferent volley reaches those neurons, the moments of the generated elementary current dipoles will add up in an optimal manner because of their parallel distribution, and the module of the resulting equivalent dipole will be considerably large. The field generated at a distance is therefore sufficiently large to produce effects measurable by an EEG [LOP 82]. These kinds of fields are known as "open fields". The most important part of an EEG recording is therefore produced by the cerebral cortex's pyramidal neurons. Even though this fact may seem discouraging, it is important to note that those neurons are very numerous and that they represent the output pathway for cortical structures. EEG is thus sensitive to the entry (PSP) of output neurons of the cerebral cortex, which is of interest.

Even though cortical pyramidal neurons are the main EEG generators, they are not the only ones. Other cortical neurons also have predominant orientations and may therefore also generate open fields. In the same manner, some groups of subcortical neurons, since they are characterized by asymmetric dendritic arborizations and since they have predominant orientations, can also generate activities measurable by EEG [LOR 47].

Let us come back to the cortical pyramidal neurons. The cerebral cortex is folded; it contains sulci and fissures (that is deep sulci). If the two banks of a sulcus are activated at the same time, the moments of the equivalent dipoles accounting for their activities are oriented in opposite directions. If their modules are of comparable sizes, they tend to cancel each other out and the activity of the two banks of the sulcus does not contribute significantly to the EEG.

3.3.2. *Effect of synapse position on the polarity of EEG*

Let us consider a structure located on a part of the cortex parallel to the surface of the scalp (gyrus). Its activation generates a current dipole perpendicular to the scalp (this is known as a "radial" dipole). Let us suppose that the activation of this area results from an excitatory afferent volley reaching the outermost layers of the cortex, that is to say the distal region of pyramidal neurons' apical dendrites (like in the neuron in Figure 3.1 on the left). The negative pole of the elementary current dipoles will be superficial and the negative pole of the equivalent dipole will be oriented toward the scalp. An EEG electrode placed above this structure will therefore record a negative deflection. If, on the other hand, the afferent volley is inhibitory, it will produce point current sources at the synapses and distributed sinks along the rest of the dendrites and cell bodies. The generated dipole's orientation will be the opposite of that in the previous example, and its positive pole will be oriented toward the scalp. An electrode placed above this structure will record a positive deflection.

Let us now suppose that the activation of this same cortical area results from an excitatory afferent volley in the deep layers, close to the neuron's soma (like the neuron in Figure 3.1 on the right). The negative pole being deep, it is the positive pole that will be directed toward the scalp and an electrode placed above it will record a positive deflection, as in the case of a superficial inhibition. In the same way, the effect of a deep inhibitory volley will be recorded by an EEG electrode as a negative deflection, such as a superficial excitatory volley.

Therefore, without specific previous knowledge regarding the generators responsible for variations in the measured potential, it is not possible to decide whether a negative wave recorded on an EEG corresponds to a superficial excitation or to a deep inhibition; nor to know if a positive wave represents a superficial inhibition or a deep excitation. However, it is certain that if a negative wave is recorded in an experimental condition and a positive wave in another condition, both in the same latency range, the negative and positive waves do indeed represent two different physiological phenomena.

3.3.3. *Effect of active areas' position*

The maximum value and the form of potentials recorded by EEG electrodes depend on the generators' position.

Let us take the expression for the potential generated by the activation of a synapse in the simplest scenario, which is the case described by the formula [3.1] above. We can see there that potential decreases proportionally with the square of the distance between the point where measurement is taken and the generator. This explains why the cortical generators contribute most to activities captured by EEG, since they are closest to the surface. This being said, there are also deep cortical generators, like the insula, the medial temporal cortex or the calcarine sulcus, whose contribution to EEG recordings is much attenuated due to their distance from the surface. This makes it quite difficult to study them with EEG. Moreover, the same formula shows that the potential generated by a dipole also depends on its orientation with respect to the recording point; indeed, it is proportional to the cosine of the angle formed by the dipole and the segment connecting its midpoint and the point where measurement is made. Since $\cos(0) = 1$ and $\cos(\pi) = -1$, the potential is greatest (in absolute value) right above the dipole if it is radial; that is to say, if it accounts for the activation of one part of the cortex located on a gyrus. Let us now consider the activation of a portion of the cortex located on the bank of a sulcus. The equivalent dipole that accounts for it is still perpendicular to the cortical surface but it is no longer right above the scalp: it is inclined in such a way that it is parallel to the scalp's surface (this is called a tangential dipole). The expression of the potential created by such a dipole shows, in a counterintuitive manner, that an electrode placed directly above the activated structure records nothing. Indeed, the angle formed by the dipole and the segment joining its midpoint and the point where measurement is taken is a right angle; since $\cos(\pi/2) = 0$, the potential measured by an electrode perpendicular to the generator is equal to zero. For a tangential dipole, the maximum and minimum potentials are located at a distance from the point right above the dipole, along the line that, on the surface, is parallel to the dipole's direction. Indeed, a distance away from the point right above the dipole, the angle diverges from $\pi/2$, which increases or decreases the value of measured potential. But, since measurements are carried out on the surface, opening the angle is accompanied by a distancing from the generator. The maximum potential amplitude generated by a tangential dipole is therefore lower than the maximum amplitude generated by a radial dipole with the same moment. We must also add that since tangential dipoles are located in sulci or fissures, they are usually deeper than radial dipoles.

3.4. The influence of conductive media

3.4.1. *Influence of glial cells*

Just like that of neurons, the membrane of glial cells (GC) is electrically charged. Their resting potential depends on the existence of a transmembrane transport of potassium, on the one hand, and on the existence of selective leak channels for potassium, on the other hand. As with neurons, their membrane potential can vary, but those variations are very slow. Given that there are at least as many GCs as neurons in humans, it is worthwhile considering their contribution to slow EEG variations, but answering that question is not easy. Variations of GC membrane potentials depend, as for neurons, on transmembrane ion currents. Unlike synapses, these ion currents are not as concentrated around a single point, although they can be fairly localized. These primary currents therefore produce secondary currents in the extracellular space and, for that reason, they contribute to slow variations in local field potentials [AMZ 00]. In the retina and the cerebellum, there are GCs with regularly organized, long extensions. If such glial arrangements exist into adulthood in the cortex, they could contribute to slow variations in EEG [SOM 75]. Moreover, intracellular compartments of adjacent GCs are connected by permeable gap junctions [GUT 81]. If incoming potassium currents located in a GC population cross their membranes, intracellular diffusion currents can propagate to neighboring cells through the gap junctions and generate distributed current sources, especially if the membrane's resistance to current leaks is high (note that GCs' resistance is rather low). Furthermore, the population's geometry will also determine whether sources' barycenter and sink coincide or not and, thereby, if variations in the GCs membrane's potential significantly contribute to slow EEG variations. It seems to us that it is difficult to answer this question at present.

However, when neurons are intensely activated, they tend to release potassium in the extracellular milieu. GCs have potassium channels and can absorb a part of the excess extracellular potassium. These potassium displacements are accompanied by water displacements that involve a swelling of the GCs and a reduction of the extracellular compartment's volume [AMZ 00]. GCs can therefore modify the composition and the volume of the extracellular space, which modifies its conductance. And yet, we have seen in the expression of the potential created by a dipole that it is inversely proportional to the medium's conductivity. Through their action on the extracellular medium, GCs can therefore bring about variations in potential, even at a distance, but those effects are not directly due to their membrane potential; they are more indirect via the effect of variations in the conductivity of the extracellular medium on the field generated by neural activity [BIR 90].

3.4.2. *Influence of skull bones*

We have up to now omitted the fact that the formula [3.1] provided above refers to a homogeneous medium. But the media through which currents flow before reaching an EEG electrode are not homogeneous at all. Currents must flow through the brain tissue, cerebrospinal fluid, the meninges and, most importantly, the bone before they can reach the scalp. They therefore encounter different types of electrical diopters, especially when they flow through the bone, which is extremely resistant. This does not contradict what we have presented thus far, but it does add some further effects.

Before reaching the scalp, currents will come up against the resistant wall formed by the bone and will tend to follow the paths of least resistance. They are thus more prone to take parallel than perpendicular directions to the bone. This produces a diffusion of currents when they cross the bone and this diffusion causes a spread of potential distributions toward the surface of the scalp, thus "blurring" the image of the distribution. This flow, which is introduced by diffusion, produces a mixing effect: activities coming from generators activated at the same time, even when they are far from one another, can partially or totally be combined, producing a sort of "spatial averaging". This averaging represents an important challenge to the interpretation of EEG phenomena.

For example, let us imagine that two different generators are activated in the same latency range. They run the risk of producing a single wave because their effects are combined in the surface. This uniqueness can make it look like a unique physiological phenomenon has been recorded and thus lead to a false interpretation of the data, even without thinking about the generators' locations. "Disentangling" the respective contributions of different generators to recorded surface activities is an important challenge for physiologists, but this question is beyond the scope of this chapter.

3.5. Conclusions

We have just seen that even though EEG only records a part of electrical brain activity, it is very sensitive to dendritic activity from an essential class of neurons in the cerebral cortex: pyramidal neurons. These represent the majority of neurons and a subset of them is responsible for the majority of cortical outputs, either toward other cortical areas or toward subcortical structures. The information provided by EEG recordings is therefore of great interest for medicine and neuroscience research.

The primary cellular mechanisms that produce currents captured by EEG are due to the opening of ion channels. Analogous mechanisms are responsible for the activity of other excitable tissues, such as striated skeletal muscle (electromyogram or EMG) or heart muscle (electrocardiogram or ECG). ECG and EMG activities are much larger than EEG and can contaminate recordings; they therefore constitute an important source of biological artifacts, which should be reduced or eliminated as much as possible in order to reliably measure EEG activity in the framework of brain–computer interfaces.

3.6. Bibliography

[AMZ 00] AMZICA F., STERIADE M., "Neuronal and glial membrane potentials during sleep and paroxysmal oscillations in the neocortex", *Journal of Neuroscience*, vol. 20, pp. 6648–6665, 2000.

[ADR 34] ADRIAN E.D., MATTHEWS B.H.C., "The Berger rhythm: potential changes from the occipital lobes in man" *Brain*, vol. 57, pp. 355–385, 1934.

[BER 29] BERGER H., "Über Elektrenkephalogramm des Menschen" *Archiv für Psychiatrie und Nervenkrankheiten*, vol. 87, pp. 527–570, 1929.

[BUZ 12] BUZSÁKI G., ANASTASSIOU C.A., KOCH C. "The origin of extracellular fields and currents-EEG, ECoG, LFP and spikes", *Nature Reviews Neuroscience*, vol. 13, pp. 407– 420, 2012.

[BIR 90] BIRBAUMER N., ELBERT T., CANAVAN A. *et al.* "Slow potentials of the cerebral cortex and behavior", *Physiological Physiological Reviews*, vol. 70, pp. 1–41, 1990.

[GUT 81] GUTNICK M.J., CONNORS B.W., RANSOM B.R., "Dye-coupling between glial cells in the guinea pig neocortical slice", *Brain Research*, vol. 213, pp. 486–492, 1981.

[HEL 99] HELMCHEN F, SVOBODA K., DENK W. *et al.*, "In vivo dendritic calcium dynamics in deep-layer cortical pyramidal neurons", *Nature Neuroscience*, vol. 2, pp. 989–996, 1999.

[HOT 80] HOTSON J.R., PRINCE D.A., "A calcium-activated hyperpolarization follows repetitive firing in hippocampal neurons", *Journal of Neurophysiology*, vol. 43, pp. 409–419, 1980.

[KAN 00] KANDEL J., SCHWARTZ J., JESSELL T., *Principles of Neural Science*, 4th ed., McGraw-Hill, New York, 2000.

[LLI 88] LLINÁS R.R., "The intrinsic electrophysiological properties of mammalian neurons: insights into central nervous system function", *Science*, vol. 242, pp. 1654–1664, 1988.

[LLI 06] LLINÁS, R.R., STERIADE M., "Bursting of thalamic neurons and states of vigilance", *Journal of Neurophysiology*, vol. 95, pp. 3297–3308, 2006.

[LLI 07] LLINÁS R.R., CHOI S., URBANO F.J. *et al.*, "γ-Band deficiency and abnormal thalamocortical activity in P/Q-type channel mutant mice", *Proceedings of the National Academy of Sciences*, vol. 104, pp. 17819–17824, 2007.

[LOP 82] LOPES DA SILVA F., VAN ROTTERDAM A., "Biophysical aspects of EEG and MEG generation", in NIEDERMEYER E., LOPES DA SILVA F., (eds), *Electroencephalography: Basic Principles, Clinical Applications and Related Fields*, Lippincott Williams & Wilkins, Baltimore, pp. 15–26, 1982.

[LOR 47] LORENTE DE NÓ R., "Action potential of the motoneuron of the hypoglossus nucleus", *Journal of Cellular and Comparative Physiology*, vol. 29, pp. 207–287, 1947.

[PER 07] PERNIER J., *Electro et magnéto encéphalographie. Biophysique, techniques et méthodes*, Hermes-Lavoisier, Paris, 2007.

[SOM 75] SOMJEN G.G., "Electrophysiology of neuroglia", *Annual Review of Physiology*, vol. 37, pp. 163–190, 1975.

[STE 88] STERIADE M. LLINÁS R.R., "The functional states of the thalamus and the associated neuronal interplay", *Physiological Reviews*, vol. 68, pp. 649–742, 1988.

4

Physiological Markers for Controlling Active and Reactive BCIs

4.1. Introduction

Despite recent progress achieved in the last few years in the field of functional imaging, identifying significant correlations between a mental task carried out by an individual and measurements of physiological variations in brain activity remains a complex problem. On the one hand, if we suppose that a given mental state is characterized by a particular signal, it is impossible in practice to specifically measure it without interference from the variety of other, non-correlated signals produced by the brain's basic activity. On the other hand, even very specific mental activity generates highly non-stationary signals – this is to say signals whose properties vary in time – due to the fact that they depend on several factors related to the individual's general state of being at the time when measurement is taken. For this reason, in order for the communication channel between the user and the system created by a BCI to be both reliable and well functioning, it is necessary to carefully select the mental tasks that the user performs, as well as the techniques employed to measure variations in his or her brain activity. The objective is to extract a specific activity devoted to communication from the "noisy background" produced by the brain when it is alert.

Chapter written by François CABESTAING and Philippe DERAMBURE.

We will focus here on *active* and *reactive* BCIs, which make it possible to establish a direct connection between users and computers in which the user sends messages or commands to the computer. Passive BCIs, in which the connection is not primarily devoted to communication but rather to analyzing the user's mental state, will be studied in the following chapter.

In an active BCI (see Figure 4.1(a)), the user brings about the communication process, which he or she controls by modifying his or her mental state at will. The interface continuously analyzes those variations and decides it has received a command when it identifies a specific kind of mental activity. In a reactive BCI (see Figure 4.1(b)), the interface emits visual, auditory or tactile stimuli to the user to which he or she may or may not pay special attention. The interface decides that a command has been emitted by the user when it detects a specific response from him or her to one or several stimuli.

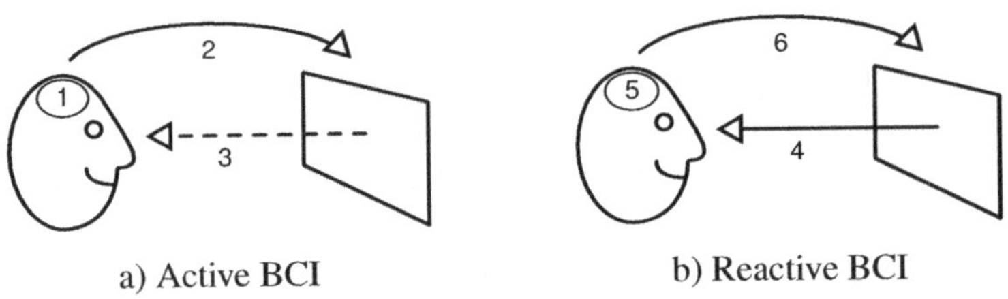

Figure 4.1. *Active versus reactive BCI. a) In an active BCI, the user spontaneously modifies his or her mental state (1). The system detects this variation (2). It then possibly sends informative feedback (3). b) In a reactive BCI, the system stimulates the user (4), whose brain reacts to certain stimuli (5), thus producing a modification in mental state eventually detected by the system (6)*

Choosing between these two approaches, and therefore between given categories of mental tasks, mainly depends on the envisioned BCI application. If the goal is to provide the user with an augmentative and alternative communication device, it is possible to consider using either approach. For example, if the aim is to command a virtual keyboard, an active approach can consist of translating the markers of a kind of spontaneous brain activity into control signals for a mouse pointer [LEE 13], whereas a reactive approach may consist of scanning through the virtual keyboard's keys by successively highlighting each one of them and then asking the user to react to the stimulus corresponding to the highlighted key he or she wants to select [FAR 88]. On the other hand, when the BCI's objective is to control an effector with several degrees of freedom – for example a prosthesis or an exoskeleton – the reactive approach is ill suited for the task.

Precisely defining a mental task is not enough to have a *marker* of identifiable brain activity that is sufficiently robust for a BCI. In order to be used as a marker, it must be possible to locate a variation in a physiological signal in space, that is to correlate it with a specific phenomenon occurring in a given location of the brain and/or of time, which is to say that it can be associated with the instant in which the user modifies his or her mental state [RAM 06]. More generally, it can be said that very specific successions in activations and deactivations of neuron assemblies, which is characteristic of large-scale integration of information by the different brain regions, represent markers that can be employed by a BCI [KRU 12]. It is also important for the user to be able to exert *voluntary control* on one or several characteristics of a physiological signal in order for it to constitute a marker. The capacity to control a physiological signal can be innate or attained through a learning process for the most part through operant conditioning techniques [BIR 13].

In order for a variation in brain activity to be used as a marker for a BCI, it is also necessary for it to be convertible into a signal through a measurement device. A key characteristic for these devices is the resolution they provide in spatial and temporal terms. Good spatial resolution makes it possible to precisely locate the brain region in which activity variations take place. Good temporal resolution makes it possible to precisely correlate variations in physiological activity with those in the user's mental state. Furthermore, if the BCI is meant to be used in *ecological* conditions, this is to say outside the very limited space of a laboratory or a medicine clinic, it is important to also take into account possible blockage of the measuring device, as well as its ease of implementation, fragility, etc.

BCIs are also categorized according to the degree of *invasiveness* they require in order to collect brain measurements. In invasive BCIs, one or several electrodes are implanted in a surgical procedure, whereas in non-invasive BCIs, brain activity measurements – which are here known as surface measurements – are recorded through an external device. In the case of invasive measurements, electrodes are placed as close as possible to the electrophysiological sources, and the measured electrical signals do not cross the skull, muscles or skin. For this reason, measurements reflect the sources' activity more accurately, which most importantly makes it possible to increase spatial resolution. Moreover, the signal to noise ratio is higher because artifacts generated by the eyes' and jaws' muscular activity can be avoided. On the other hand, invasive techniques require precise knowledge of the location of electrophysiological sources in order to optimize implantation of the electrodes before surgical procedures.

The vast majority of measurement techniques currently available have been employed more or less thoroughly with the objective of implementing BCIs [BIR 06]. Physiological activities, devices enabling their measurement, spatial and temporal resolution levels provided by those devices and the characteristics of the corresponding BCIs are listed in Table 4.1. Note that electrophysiological activity measurements are all capable of being used in an ecological BCI context. On the other hand, although its spatial and temporal resolution are adequate, magnetoencephalography equipment is burdensome, expensive and must be installed in a perfectly isolated environment in terms of electromagnetic disturbances. Functional imaging techniques, which detect variations in brain metabolism through blood oxygen level dependent (BOLD) signals, are characterized by great temporal imprecision related to that signal's dynamic. Furthermore, these measurement devices are currently very bulky, even if it is reasonable to imagine that fNIRS spectroscopy systems, which are inexpensive, of manageable dimensions, and have a good spatiotemporal resolution, will become available in a few years [ZEP 14]. Several authors have thought to overcome the limitations related to using a single measurement technique by employing several of them, thus leading to so-called *hybrid* BCIs, although the description of those techniques goes beyond the scope of this chapter.

Some experiments have been carried out to verify the possibility of using electrical signals gathered on a unique neuron or nerve fiber. For example, Kennedy *et al.* recount their use of neurotrophic electrodes implanted in patients suffering from amyotrophic lateral sclerosis [KEN 04]. Many of those patients have successfully managed to control a spelling interface by modulating the action potential discharge rhythm emitted by a recorded neuron. Very few authors have reported the use of a grid of electrodes chronically implanted in a subject's motor cortex in order to implement a BCI. The most recent experiment to have been attempted was carried out by Hochberg *et al.* on two quadriplegic subjects, within the framework of the

BrainGate 2 clinical project [HOC 12]. Signals gathered by a grid of 96 electrodes were processed in order to bring to light unitary action potentials from a neuron ensemble in the primary motor cortex area controlling the hand. Those signals were then decoded in order to calculate the speed of movement intentions in real time, which were in turn used to control a robotic arm.

Activity, measurement technique	*Spatial res.*	*Temporal res.*	*Invasive*	*Ecological*
Action potential, electrode on a neuron connected to nerve fiber	Excellent	Excellent	Yes	Yes
Local field potential	Good	Excellent	Yes	Yes
Electrocorticogram, epidural or sub dural electrode grid	Adequate	Excellent	Yes	Yes
Surface electroencephalogram	Bad	Excellent	No	Yes
Magnetoencephalogram	Adequate	Excellent	No	No
BOLD signal measured by functional magnetic resonance imaging (fMRI)	Adequate	Bad	No	No
BOLD signal measured by funtional near infrared spectroscopy (fNIRS)	Bad	Bad	No	No (*)

Table 4.1. *Brain activity signals already used in a BCI [BIR 06], with associated measurement techniques, spatial and temporal resolutions, invasiveness and ecological characteristics. BOLD = Blood-Oxygen-Level Dependent; fMRI = functional Magnetic Resonance Imaging; fNIRS = functional Near Infrared Spectroscopy*

In this chapter, we will describe electrophysiological activities whose variations, when recorded by a surface electroencephalogram (EEG), are analyzed in order to extract a marker. It is important to note that these are the same activities that are studied in invasive interfaces, where signals are recorded by epidural or subdural electrode grids (electrocorticogram), with the slight difference that spatial resolution and the signal to noise ratio are higher in this case. Variation in EEG signals can be extracted either directly in the spatiotemporal domain, or after a frequency transformation. Spatiotemporal analysis is used to detect an event-related potential (ERP), for which there is a correlation between the phase of oscillations appearing as a response to an event and the instant at which it occurs. Increases or decreases as a function of time are studied at different spatial locations, with the instant at which the event occurs taken as the origin. On the other hand, in order to detect an *induced* activity, given that the phase is not locked, it is only possible to analyze variations in oscillation amplitude after the event's occurrence.

In an active BCI, events that bring about a modification in the individual's mental state are *endogenous*, with the individual spontaneously deciding to carry out a specific mental task. Instead, in a reactive BCI, events that can bring about a modification in the individual's mental state are called *exogenous*, which is to say that they are caused by sensory stimulation.

4.2. Markers that enable active interface control

In this section, we first describe a marker for an activity evoked by an endogenous event that has been widely used to implement BCIs, namely slow variations in average cortical potential. We also present a marker related to the realization of a motor imaging task, namely bereitschaftspotential (BP) or readiness potential.

These first two markers are extracted through a spatiotemporal analysis of EEG signals. Next, we describe markers that conform desynchronizations and synchronizations related to an event, especially desynchronization of beta ($f = 13 - 30$ Hz) and mu ($f = 8 - 12$ Hz) waves before and during motor imaging, and the resynchronization of beta waves afterwards. This phenomenon is known as a "beta rebound". In this case, the markers are extracted by an analysis of EEG signal variations in terms of frequency.

4.2.1. *Spatiotemporal variations in potential*

4.2.1.1. *Slow variations of average cortical potential*

A slow evolution of average cortical potential (slow cortical potential [SCP]) is categorized according to the sign of its variation – either positive or negative – with respect to the average or baseline level, as measured on an interval of time in which the user is considered at rest. The measured potential corresponds to the dendrites' level of depolarization in the upper cortex, whose variations are caused by afferent, intracortical or thalamo cortical influxes advancing toward layers I and II of the neocortex. Negative SCPs are the result of very slow synchronized excitatory postsynaptic potentials emitted by the apical dendrites from the pyramidal neurons. Positive SCPs are the result of a decrease in those same potentials, an inhibitive activity in the interneurons, or an excitatory influx coming from the cell bodies in layers IV and V [BIR 00].

The spatial position at which the maximum variations in average cortical potential occurs varies among individuals [HIN 04]. Recordings carried out on several EEG channels during a visual feedback experiment for BCI learning have shown that SCPs with the greatest amplitude tend to appear at or around the vertex, but that variations occurring in a large spatial extension can also be recorded in some subjects [HIN 05].

Slow average cortical potential variations can constitute a useful marker for BCIs because (1) they are easy to detect by comparing the level of average electrical potential at the instant in question – which is measured by using one or several surface EEG electrodes at the instant – with the baseline potential; (2) they manifest themselves in an innate manner due to the fact that they are the product of changes in excitation levels of a cortical network related to the preparation of a cognitive or motor task; and (3) a user can learn to voluntarily modulate these variations in order to increase their amplitude by employing operant conditioning techniques [BIR 00].

The learning technique implemented by Birmaumer to this end is described in detain in section 3.3.1 of Chapter 3 of Volume [CLE 16]. This marker is of interest because SCPs are measurable even for patients suffering from major brain injury, like those caused by an ALS.

4.2.1.2. *BP or readiness potential*

Discovered over 50 years ago, readiness potential manifests itself in two successive phases through a reduction of cortical potential that is at first slow and then fast. BP begins between 1 s and 1.5 s before the execution of a voluntary movement [KOR 65]. Its first component (EBP, early BP) is more visible in the centromidline, whereas the second component (LBP, late BP), which starts about 400 ms before movement, has its maximum amplitude right above the primary motor cortex (see Figure 4.2). The generally accepted hypothesis is that EBP begins in the presupplementary motor area (pre-SMA) without a specific location and in the SMA displaying a somatotopic organization, and then bilaterally continues on to the lateral premotor cortex. On the other hand, LBP is generated specifically in the contralateral primary motor cortex (M1) and the lateral premotor cortex following a precise somatotopy [SHI 06].

BP also manifests itself when the subject observes movement being carried out by another person, when he or she can predict another person's movement, and finally when he or she imagines performing that movement. It is this last property that makes it possible to consider BP an electrophysiological marker usable in an active BCI. Either on its own or in conjunction with other markers, BP has been used in several BCI implementations in order to predict when a subject has the intention of carrying out a movement, the direction of that movement and even the limb that will be involved in that movement [AHM 13].

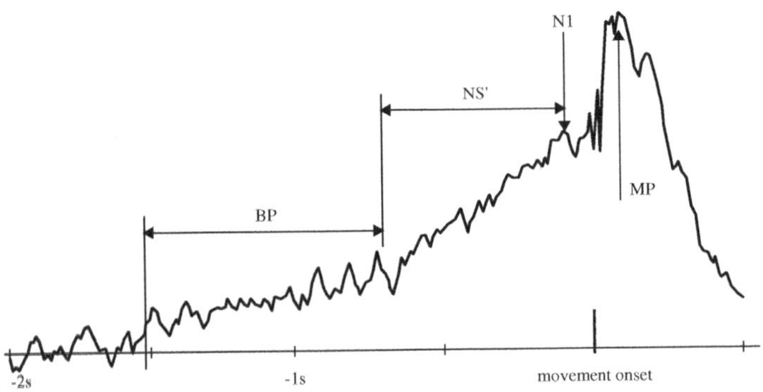

Figure 4.2. *Temporal variation in potential before movement. BP, bereitschaftspotential; NS', negative slope; MP, motor potential*

4.2.2. *Spatiotemporal wave variations*

Some authors consider ERPs to be the superposition of several responses produced by the event, leading to a measurable macroscopic variation. Some other authors consider them a simultaneous zero-reset of a variety of brain waves, which in fact brings about variations in potential that are clearly apparent in the temporal domain [SAU 07]. When the event is endogenous, resynchronization of the phases is not as marked, and its effect on EEG wave oscillations can only be extracted through frequency analysis.

The most frequently used BCI markers of an activity induced by an endogenous event are synchronizations and desynchronizations related to the event (event-related synchronization (ERS), event-related desynchronization (ERD)). In the case of an ERS, variations in activity are manifested in a temporal increase – with respect to its average value – of oscillation amplitude in a specific frequency range. In the case of an ERD, the opposite result holds: a decrease in the amplitude of those same oscillations can be detected. According to the type and complexity of the endogenous event inducing the variation in brain activity, the spatial location of ERD/ERS and their latency – that is to say the time between the instant they appear and the moment the event was produced – are very variable.

In active BCIs employing ERD/ERS, the endogenous event corresponds in practice to the user beginning a specific mental task. For example, carrying out a mental calculation, which involves the use of working memory, produces an ERS in the gamma range ($f > 25$ Hz) in the prefrontal dorsolateral cortex [RAM 06]. Among the mental tasks that have already been the subject of active BCI experiments employing an ERD/ERS, we can find the following [FRI 12, DEL 02]: rotating an object, cube or more complex shape; associating words, for example those beginning with a given letter; auditory imagination; mental navigation of a known space; imagining faces of known people; motor imaging.

Without a doubt, evidencing ERDs/ERSs produced by a motor imaging task has been the object of the largest number of research devoted to active BCI development. This can be explained by the fact that ERDs/ERSs are produced for this mental task in frequency ranges – mu ($f = 8 - 12$ Hz) and beta ($f = 13 - 30$ Hz) – that can be easily measured by a surface EEG, unlike ERDs/ERSs produced for oscillations at higher frequencies. Moreover, movement-correlated mu rhythm ERDs/ERSs can easily be spatially located given that maximum amplitude variations are recorded on EEG electrodes located right above the primary motor areas in the precentral gyrus [DER 99].

Figure 4.3 represents an EEG signal's power spectrogram as recorded when focusing on the motor area during movement. On this spectrogram, desynchronization (ERD) and then resynchronization (ERS) of the rhythms related to movement execution can be seen clearly. Neuper and Pfurtscheller have shown that the same power variations occur in EEG signals when movement is simply imagined rather than actually carried out [NEU 98]. A motor imaging mental task therefore normally involves a desynchronization of beta and mu rhythms, followed by a resynchronization of the beta rhythm, which can be detected in order to serve as usable active BCI markers.

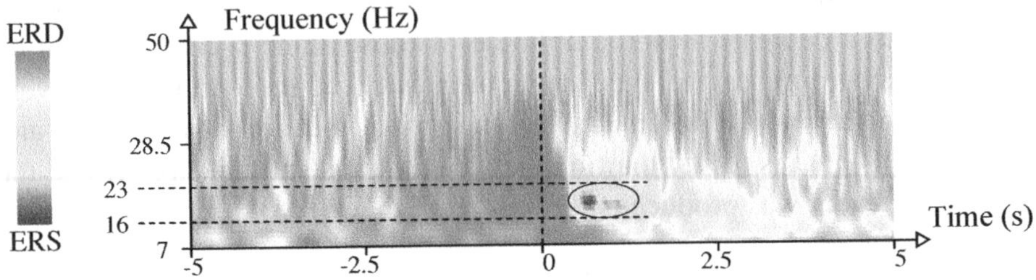

Figure 4.3. *Desynchronization and synchronization related to the event. Spectrogram (time-frequency representation) of an EEG signal recorded directly above the motor area during movement. Event-related desynchronization (ERD) begins both in the mu (8–12 Hz) and beta (13–30 Hz) ranges. ERD in the mu range is produced for about 1 s after the end of movement. On the other hand, only 500 ms after the end of movement an event-related synchronization appears in the beta range. This "beta rebound" is located in the figure's ellipse. For a color version of this figure, see www.iste.co.uk/clerc/interfaces1.zip*

However, imagining movement without actually carrying it out is not a trivial mental task, and the cognitive strategies employed by subjects that perform it are in practice very variable, even in the case of people suffering from paralysis [MUL 07]. Several studies have shown that short duration movement imagination is a strategy that produces straightforward desynchronizations in beta and mu rhythms at the beginning of the task [MOR 08, JEO 11]. On the other hand, when the subject imagines movements for a longer period of time or for a variable length of time, ERDs are less marked at the beginning of the task [JEO 11]. In this case, the motor imaging task is more easily captured by the resynchronization of mu and beta rhythms that appear in EEG signals when it ends.

A subject's capacity to voluntarily modify the amplitude of his or her sensorimotor rhythms with the purpose of controlling an active BCI can also be improved by an operant conditioning technique. Wolpaw *et al.* were the first to relate this approach, initially for controlling the vertical position of a cursor (just one degree of freedom [WOL 91]), and then on two screen dimensions (two degrees of freedom [WOL 94]). In this work, BCI control signals are determined by measuring the power of EEG signals recorded directly above the primary motor area, in the frequency range corresponding to the mu rhythm, while respecting motor somatotopy.

4.3. Markers that make it possible to control reactive interfaces

In cases qualified as reactive, a BCI sends the user repetitive stimulations whose spatial and/or temporal characteristics encode two or more options which he or she must choose from. The user therefore has the task of focusing, or not focusing, his or her attention on the stimulations associated with the option he or she wishes to select in order to communicate with the interface. The very perception of stimulation produces a potential, which is qualified as exogenous. If the perception moreover brings about a specific cognitive reaction in the subject, other potentials can also be produced later, like P300. Among the exogenous potentials that are employed as markers in BCIs, we can mostly find stationary sensory evoked potentials (SEPs), which are produced by a visual, auditive or tactile stimulation and which are locked in phase with this stimulation. P300 is the cognitive evoked potential brought about by a stimulation that is most commonly used as a marker in BCIs. We will therefore describe it in detail.

4.3.1. *Sensory evoked potentials*

SEPs appear in EEG signals only when the subject receives sensory stimulation of some type. In this sense, they are different from spontaneous potentials and ERPs that are not related to perception, for example the spontaneous decision to perform a given movement. SEPs are locked in phase with stimuli and can be determined with a high degree of precision. This property makes it possible to easily extract them by repeating the simulation a large number of times and averaging EEG readings, but also, in the case of a single reading, to acquire good *a priori* knowledge of characteristic variations that are being studied in the signals. In order to use an SEP as a marker in a BCI, it is in practice necessary for it to be identified and located in the EEG signals by using a very limited number of simulation repetitions, so as to obtain rapid communication.

4.3.1.1. *Visual evoked potential*

Just after the perception of a visual stimulus with a very short duration – generally a flash of light – it is possible to record several SEPs of different latencies directly above the primary visual area. In this case, the visual evoked potentials (VEPs) with the highest amplitude are N2, with a latency of 90 ms, and P2, with a latency of 120 ms. When stimulation is produced by a mechanism whose luminescence remains constant, but whose contrast is suddenly modified – for example by changing the black and white tiles of a checkerboard – the most marked VEPs are N75, P100 and N135 potentials [VIA 10]. In those simulation conditions, that is considering a response to a single stimulus, it is extremely difficult to extract VEPs in EEG signals in a single attempt. For this reason, their use as BCI markers is unlikely [ZHU 10].

On the other hand, however, when simulations are repetitive, a permanent functioning state is established in the visual perception chain, leading to the emergence of steady-state visually evoked potentials (SSVEPs). SSVEPs, which were first studied in humans by Regan in 1966 [REG 66], are characterized by their amplitude and phase. The phenomenon's phase constitutes the reference stimulus. These characteristics are generally measured through a frequency transformation of the EEG content, much like that used to analyze cortical rhythms. Even if the question has not been definitively determined, it is considered that the stimulation frequencies that allow for SSVEP evocation are between 3 and 40 Hz [VIA 10]. Below 3 Hz, the stimuli's cadence is insufficient for a steady-state protocol to set up. Brain responses therefore become equivalent to a succession of unitary responses. Beyond 40 Hz, the SSVEP's amplitude becomes too small to be detected and processed, especially from surface EEG signals. Figure 4.2 provides an example of an SSVEP measured by an occipital electrode placed in Oz, with the reference taken at the vertex.

SSVEPs are interesting markers for BCI implementation because (1) they are easy to extract with basic processing of EEG signals, for example narrowband filtering centered on the stimulus' frequency; (2) their characteristics are relatively independent of the subject, with the condition of specifically selecting electrode positions providing maximum amplitude; and (3) they are little affected by muscular artifacts such as blinking and facial muscles' EMG [PER 03].

In the majority of BCIs employing SSVEPs, the subject chooses an option by directing his or her gaze toward one of the available targets, each flashing at a different frequency. For this reason, the resulting BCI is qualified as *dependent*, since the subject must at least be in control of some residual movement allowing him or her to change the direction of his or her gaze. Recent studies have talked about *independent* BCIs employing SSVEP evoked by spatially combined stimulations, for example a checkerboard whose two types of tiles are colored differently and flash at two different frequencies [ALL 08, LES 14]. In those BCI arrangements, the SSVEP amplitude is determined by the attention that the subject pays to one or another of the stimulations, and not by a change in the gaze direction.

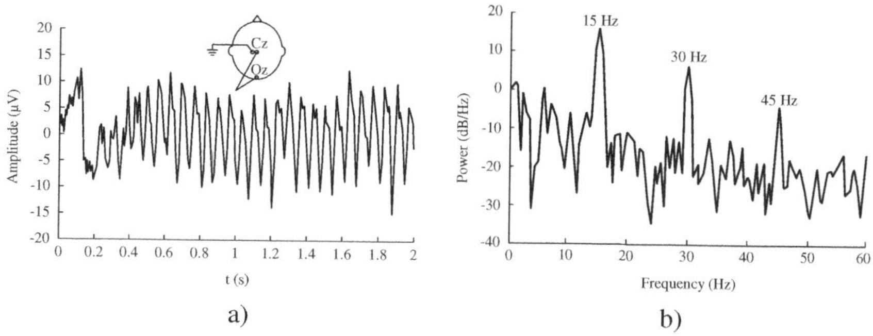

Figure 4.4. *EEG measured on a bipolar derivation on positions Oz and Cz during visual stimulation at a frequency of 15 Hz. a) Signal averaged on 10 experiments synchronized at the beginning of the stimulus. The transitory phase is clearly visible before the emergence of the steady-state protocol. b) Power spectrum for that same signal, showing a peak for the fundamental frequency – which is identical to that of the stimulus – and several other peaks for its harmonics. Figure reproduced from [ZHU 10]*

4.3.1.2. *Other steady-state potentials: steady-state auditory evoked potential, auditory steady-state response and steady-state somatosensory evoked potential*

Repetitive stimulation of a sensory modality other than vision also makes it possible to evoke steady-state SEPs. Somatosensory evoked potentials can therefore also be detected for repetitive tactile stimulation (steady-state somatosensory evoked potentials) as can auditory evoked potentials for repetitive auditive stimulation (steady-state auditory evoked potential (SSAEP), or auditory steady-state response). As for SSVEPs, these steady-state potentials can be extracted from EEG signals through narrowband filtering centered on the stimulation frequency or one of its harmonics. Indeed, the possibility of using those steady-state SEPs as markers in a reactive BCI is still being debated by the community, especially in the case of SSAEPs. The problem has to do with the fact that very little research has been able to demonstrate a user's capacity to adjust his or her SEPs by selectively focusing attention on a specific stimulus [HIL 12, MUL 06].

4.3.2. *Endogenous P300 potential*

When an individual receives an unexpected sensory stimulus, or one that is different from what he or she expected to perceive, the standard SEP is followed by one or several other, very specific evoked potentials, whose latencies are greater [DES 65, SUT 65]. If the unknown stimulus's perception captures the individual's attention, that is to say it elicits a cognitive reaction from him or her, an endogenous potential called novelty P3 or P3a appears. The latency of P3a falls somewhere between 250 and 280 ms. Its spatial distribution is frontocentral, and its amplitude is correlated with the strength of the surprise effect caused by the stimulus. However, if the initially unknown stimulus appears several times, the individual becomes accustomed to it quickly and the amplitude of P3a greatly decreases.

P3b, which is sometimes referred to as target P3, is also evoked in response to an unexpected sensory stimulus. In this case, the stimulus is known beforehand by the individual, who must be involved in a cognitive task consisting of reacting when he or she perceives that particular stimulus in a sequence of other stimuli. The latency of P3b is for the most part greater than that of P3a, between 250 and 500 ms, or greater according to the complexity of the cognitive task performed by the individual. Moreover, the spatial distribution of P3b is more posterior than that of P3a.

In some reactive BCIs, P3b endogenous evoked potential, called P300 in that case, is detected by EEG signals processing in order to determine the specific stimulus on which the user focuses his or her attention. In the "oddball" paradigm, a succession of stimuli containing at least two different types of stimuli is presented, usually at a relatively fast rate [SQU 75]. The user is asked to react each time the target stimulus appears, for example by counting the number of times it appears. The other stimuli are considered distractions (standard stimuli). The frequency with which they appear must be greater than that of the target stimuli in order for the amplitude of P300 to be large enough.

P300 is the evoked potential that has most commonly been used in reactive BCIs since Farwell and Donchin's publication [FAR 88]. It represents an especially relevant marker because (1) it can be measured in the first session of BCI use due to the fact that it results from an innate learning process; (2) it is often possible to detect it in a single test, especially when the space–time detection filter's settings are well adapted to the user; (3) the paradigms used in P300 BCIs make high-speed stimulations possible – going as high as 10 stimuli per second – which makes it possible to attain adequate communication lags.

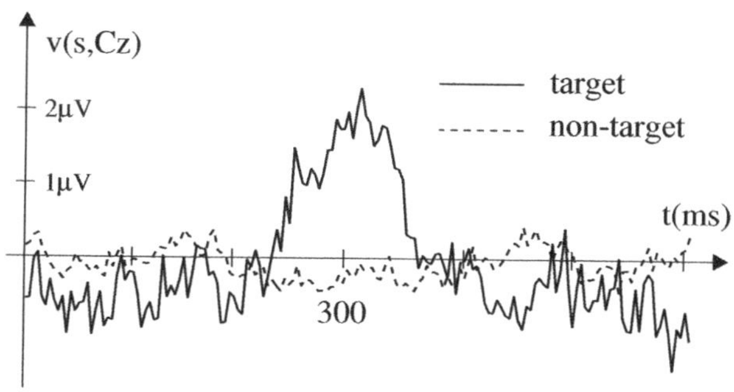

Figure 4.5. *Endogenous P3b potential. EEG readings recorded on a Cz electrode in response to target stimuli (continuous line) and to standard stimuli (dotted line), averaged over 60 recording in each case. We can observe an increase in potential in the neighborhood of 300 ms after a target stimulus*

4.4. Conclusions

In this chapter, we have described the main physiological markers that can be used in order to control an active or reactive BCI. In the case of active BCIs, we have presented the markers of an evoked activity through an endogenous event: slow variations in average cortical potential, BP, synchronizations and desynchronizations resulting from an event. In the case of reactive BCIs, we described the markers of an activity evoked by an exogenous event, in this case a stimulation emitted by the BCI: steady-state somatosensory evoked potentials, especially visual (SSVEP) and cognitive (P300) evoked potentials.

4.5. Bibliography

[AHM 13] AHMADIAN P., CAGNONI S., ASCARI L., "How capable is non-invasive EEG data of predicting the next movement? A mini review", *Frontiers in Human Neuroscience*, vol. 7, no. 124, pp. 1–7, April 2013.

[ALL 08] ALLISON B.Z., MCFARLAND D.J., SCHALK G. *et al.*, "Towards an independent brain–computer interface using steady state visual evoked potentials", *Clinical Neurophysiology*, vol. 119, no. 2, pp. 399–408, February 2008.

[BIR 00] BIRBAUMER N., KÜBLER A., GHANAYIM N. *et al.*, "The thought translation device (TTD) for completely paralyzed patients", *IEEE Transactions on Rehabilitation Engineering*, vol. 8, no. 2, pp. 190–193, June 2000.

[BIR 06] BIRBAUMER N., "Breaking the silence: brain–computer interfaces (BCI) for communication and motor control", *Psychophysiology*, vol. 43, no. 6, pp. 517–532, November 2006.

[BIR 13] BIRBAUMER N., RUIZ S., SITARAM R., "Learned regulation of brain metabolism", *Trends in Cognitive Sciences*, vol. 17, no. 6, pp. 295–302, June 2013.

[CLE 16] CLERC M., BOUGRAIN L., LOTTE F., *Brain–Computer Interfaces 2: Technology and Application*, ISTE, London and John Wiley and Sons, New York, 2016.

[DEL 02] DEL R. MILLÀN J., NO J.M. *et al.*, "A local neural classifier for the recognition of EEG patterns associated to mental tasks", *IEEE Transactions on Neural Networks*, vol. 13, no. 3, pp. 678–686, May 2002.

[DER 99] DERAMBURE P., DEFEBVRE L., BOURRIEZ J.-L. *et al.*, "Désynchronisation et synchronisation liées à l'événement. Étude de la réactivité des rythmes électrocorticaux en relation avec la planification et l'exécution du mouvement volontaire", *Neurophysiologie Clinique*, vol. 29, no. 1, pp. 53–70, February 1999.

[DES 65] DESMEDT J., DEBECKER J., MANIL J., "Mise en évidence d'un signe électrique cérébral associé à la détection par le sujet d'un stimulus sensoriel tactile", *Bulletin de l'Académie Royale de Médecine de Belgique*, vol. 5, no. 11, pp. 887–936, 1965.

[FAR 88] FARWELL L.A., DONCHIN E., "Talking off the top of your head: toward a mental prosthesis utilizing event-related brain potentials", *Electroencephalography and Clinical Neurophysiology*, vol. 70, no. 6, pp. 510–523, December 1988.

[FRI 12] FRIEDRICH E.V., SCHERER R., NEUPER C., "The effect of distinct mental strategies on classification performance for brain–computer interfaces", *International Journal of Psychophysiology*, vol. 84, no. 1, pp. 86–94, April 2012.

[HIL 12] HILL N.J., SCHÖLKOPF B., "An online brain–computer interface based on shifting attention to concurrent streams of auditory stimuli", *Journal of Neural Engineering*, vol. 9, no. 2, pp. 1–13, February 2012.

[HIN 04] HINTERBERGER T., SCHMIDT S., NEUMANN N. *et al.*, "Brain-computer communication and slow cortical potentials", *IEEE Transactions on Biomedical Engineering*, vol. 51, no. 6, pp. 1011–1018, June 2004.

[HIN 05] HINTERBERGER T., VEIT R., WILHELM B. *et al.*, "Neuronal mechanisms underlying control of a brain–computer interface", *European Journal of Neuroscience*, vol. 21, no. 11, pp. 3169–3181, June 2005.

[HOC 12] HOCHBERG L.R., BACHER D., JAROSIEWICZ B. *et al.*, "Reach and grasp by people with tetraplegia using a neurally controlled robotic arm", *Nature*, vol. 485, no. 7398, pp. 372–375, May 2012.

[JEO 11] JEON Y., NAM C.S., KIM Y.-J., *et al.*, "Event-related (De)synchronization (ERD/ERS) during motor imagery tasks: implications for brain-computer interfaces", *International Journal of Industrial Ergonomics*, vol. 41, no. 5, pp. 428–436, September 2011.

[KEN 04] KENNEDY P., KIRBY M., MOORE M. *et al.*, "Computer control using human intracortical local field potentials", *IEEE Transactions on Neural Systems and Rehabilitation Engineering*, vol. 12, no. 3, pp. 339–344, September 2004.

[KOR 65] KORNHUBER H.H., DEECKE L., "Hirnpotentialänderungen bei Willkürbewegungen und passiven Bewegungen des Menschen: Bereitschaftspotential und reafferente Potentiale", *Pflüger's Archiv für die gesamte Physiologie des Menschen und der Tiere*, vol. 284, no. 1, pp. 1–17, May 1965.

[KRU 12] KRUSIENSKI D.J., MCFARLAND D.J., WOLPAW J.R., "Value of amplitude, phase, and coherence features for a sensorimotor rhythm-based brain-computer interface", *Brain Research Bulletin*, vol. 87, no. 1, pp. 130–134, January 2012.

[LEE 13] LEEB R., PERDIKIS S., TONIN L. *et al.*, "Transferring brain–computer interfaces beyond the laboratory: successful application control for motor-disabled users", *Artificial Intelligence in Medicine*, vol. 59, no. 2, pp. 121–132, October 2013.

[LES 14] LESENFANTS D., HABBAL D., LUGO Z. *et al.*, "An independent SSVEP-based brain–computer interface in locked-in syndrome", *Journal of Neural Engineering*, vol. 11, no. 3, pp. 1–8, May 2014.

[MOR 08] MORASH V., BAI O., FURLANI S. *et al.*, "Classifying EEG signals preceding right hand, left hand, tongue, and right foot movements and motor imageries", *Clinical Neurophysiology*, vol. 119, no. 11, pp. 2570-2578, November 2008.

[MUL 06] MULLER-PUTZ G.R., SCHERER R., NEUPER C. *et al.*, "Steady-state somatosensory evoked potentials: suitable brain signals for brain-computer interfaces?", *IEEE Transactions on Neural Systems and Rehabilitation Engineering*, vol. 14, no. 1, pp. 30–37, March 2006.

[MUL 07] MULLER-PUTZ G.R., ZIMMERMANN D., GRAIMANN B. *et al.*, "Event-related beta EEG-changes during passive and attempted foot movements in paraplegic patients", *Brain Research*, vol. 1137, pp. 84–91, 2007.

[NEU 98] NEUPER C., PFURTSCHELLER G., "ERD/ERS based brain computer interface (BCI): effects of motor imagery on sensorimotor rhythms", *International Journal of Psychophysiology*, vol. 30, pp. 53–54, 1998.

[PER 03] PERLSTEIN W.M., COLE M.A., LARSON M. *et al.*, "Steady-state visual evoked potentials reveal frontally-mediated working memory activity in humans", *Neuroscience Letters*, vol. 324, no. 3, pp. 191–195, May 2003.

[RAM 06] RAMSEY N.F., VAN DE HEUVEL M.P., KHO K.H., *et al.*, "Towards human BCI applications based on cognitive brain systems: an investigation of neural signals recorded from the dorsolateral prefrontal cortex", *IEEE Transactions on Neural Systems and Rehabilitation Engineering*, vol. 14, no. 2, pp. 214–217, June 2006.

[REG 66] REGAN D., "Some characteristics of average steady-state and transient responses evoked by modulated light", *Electroencephalography and Clinical Neurophysiology*, vol. 20, no. 3, pp. 238–248, March 1966.

[SAU 07] SAUSENG P., KLIMESCH W., GRUBER W. *et al.*, "Are event-related potential components generated by phase resetting of brain oscillations? A critical discussion", *Neuroscience*, vol. 146, no. 4, pp. 1435–1444, June 2007.

[SHI 06] SHIBASAKI H., HALLETT M., "What is the Bereitschaftspotential?", *Clinical Neurophysiology*, vol. 117, no. 11, pp. 2341–2356, November 2006.

[SQU 75] SQUIRES N.K., SQUIRES K.C., HILLYARD S.A., "Two varieties of long-latency positive waves evoked by unpredictable auditory stimuli in man", *Electroencephalography and Clinical Neurophysiology*, vol. 38, no. 4, pp. 387–401, April 1975.

[SUT 65] SUTTON S., BRAREN M., ZUBIN J., *et al.*, "Evoked-potential correlates of stimulus uncertainty", *Science*, vol. 150, no. 3000, pp. 1187–1188, November 1965.

[VIA 10] VIALATTE F.-B., MAURICE M., DAUWELS J., *et al.*, "Steady-state visually evoked potentials: focus on essential paradigms and future perspectives", *Progress in Neurobiology*, vol. 90, no. 4, pp. 418–438, April 2010.

[WOL 91] WOLPAW J.R., MCFARLAND D.J., NEAT G.W., *et al.*, "An EEG-based brain–computer interface for cursor control", *Clinical Neurophysiology*, vol. 78, pp. 252–259, 1991.

[WOL 94] WOLPAW J.R., MCFARLAND D.J., "Multichannel EEG-based brain–computer communication", *Electroencephalography and Clinical Neurophysiology*, vol. 90, no. 6, pp. 444–449, June 1994.

[ZEP 14] ZEPHANIAH P.V., KIM J.G., "Recent functional near-infrared spectroscopy based brain computer interface systems: developments, applications and challenges", *Biomedical Engineering Letters*, vol. 4, no. 3, pp. 223–230, September 2014.

[ZHU 10] ZHU D., BIEGER J., MOLINA G.G., *et al.*, "A survey of stimulation methods used in SSVEP-based BCIs", *Computational Intelligence and Neuroscience*, vol. 2010, no. 702357, pp. 1–12, January 2010.

5

Neurophysiological Markers for Passive Brain–Computer Interfaces

5.1. Passive BCI and mental states

5.1.1. *Passive BCI: definition*

Recently, brain–computer interface (BCI) development has turned away from its original purpose – namely a subject controlling an effector – and it has focused instead on providing a mental state estimation tool. It is therefore common to speak of "passive" BCIs to refer to systems that no longer employ voluntarily directed brain activity aimed at controlling an effector, but that instead employ signals involuntarily generated by an individual in their task to enrich man–machine communication in an *implicit* manner in addition to monitoring operators in a variety of risky work environments (for example piloting, driving, plant surveillance) [GEO 10, PUT 10, ZAN 11]. This new tool, which brings together neuroimaging and signal processing, satisfies a need in the field of mental state monitoring and makes it possible to respond to a growing need for neuroergonomics. The general structure of a passive BCI is described by Figure 5.1. It covers brain signal acquisition, a processing chain that includes preprocessing and signal conditioning, pertinent marker extraction, translation of those markers and use of that translation to adapt the system or provide feedback. As we will see, different mental states can

Chapter written by Raphaëlle N. ROY and Jérémy FREY.

thereby be characterized and estimated by passive BCIs. This estimation is based on a variety of neurophysiological markers that we will discuss in further detail in the following.

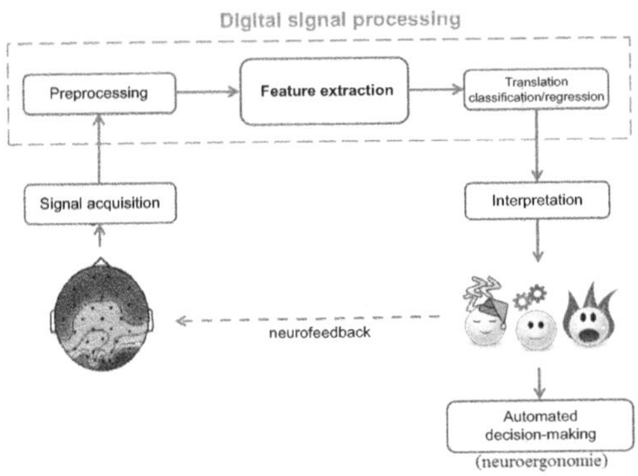

Figure 5.1. *General structure of a passive BC. For a color version of this figure, see www.iste.co.uk/clerc/interfaces1.zip*

5.1.2. *The notion of mental states*

Passive BCIs have the objective of estimating a diversity of mental states, such as the engagement of attentional and cognitive resources – which intersects with attention, mental fatigue and cognitive load – and, as well as it, error detection and the operator's emotional state, to name just a few. Each of these mental states is defined in terms of the cognitive and physiological processes it involves, as well as of behavioral performances and neurophysiological markers. In this chapter, we have focused on electroencephalographic markers of some of the major mental states in the scientific literature. These states are reported separately but it is worth noting that there are interactions between mental states, interactions that operate at the cognitive level and have repercussions at the behavioral and neurophysiological levels. These interactions remain for the most part undocumented and currently existing systems do not take them into account.

5.1.3. *General categories of neurophysiological markers*

BCIs use different kinds of markers extracted from target signals. For example, spectral markers (e.g. power in a given frequency band), temporal markers (e.g. potentials evoked by a given stimulation) and spatial markers (e.g. relative activation of some brain areas) can be used. These three types of markers are described in Chapter 4.

5.2. Cognitive load

5.2.1. *Definition*

Cognitive load can be defined as the task difficulty and the associated effort furnished by the operator to respond to the task's demands. [GEV 07]. It therefore depends on each individual's capabilities and engagement [CAI 07]. The term *cognitive load* is very general, and it is used in research that focuses on variations in task difficulty in terms of:

– the number of items to remember (short term/working memory);

– the number of tasks to be carried out at the same time (divided attention);

– perceptive difficulty (e.g. 3D visual perception);

– temporal pressure (stress).

5.2.2. *Behavioral markers*

The cognitive load's effect on behavior has been extensively explored in the literature. Higher degrees of difficulty will therefore result in a decrease in behavioral performance, whether it is an increase in reaction time or a decrease in response accuracy. For example, using Sternberg's classical paradigm, which requires participants to encode and retrieve items in working memory [STE 66], it has been shown that subjects' reaction time increases linearly with the number of items to memorize [STE 69, GOM 06].

5.2.3. *EEG markers*

From the perspective of frequency, several studies have demonstrated a modulation in the power spectral density of some frequency bands in response to changes in cognitive load levels. Therefore, during increases in cognitive load a decrease in alpha power spectral density (8–14 Hz) at centro-parietal sites (e.g. Pz) is observed along with an increase in theta (4–8 Hz), or even delta (0–4 Hz) power at centrofrontal sites (e.g. Fz) [GEV 00, MIS 06, GOM 06, HOL 09, ANT 10]. Several studies also show variations in the gamma frequency band due to modulations in task difficulty and mental load. [BER 07, OSS 11]. These variations have led to the creation of several indexes, such as the ratio of frontocentral activity in theta and parietal activity in alpha [GEV 03, HOL 09].

With respect to studies focusing on evoked potentials, they have mainly been carried out through concurrent target detection tasks. The amplitude of the P300 component would be a reliable working memory resource allocation index [KOK 01, FU 07]. Indeed, research studies show a decrease in P300 amplitude following an increase in cognitive load [NAT 81, KOK 01, HOL 09, SCH 04]. Beyond the P300 component, some authors have also brought to light a modulation of earlier neuronal components in response to variations in cognitive load. Components N1, N2 and P2 see their amplitude decrease with an increase in load [ALL 08, MIL 11]. Finally, Miller *et al.* have shown a decrease in the amplitude of the Late Positive Potential (LPP) with an increase in the task's difficulty [MIL 11].

5.2.4. *Application example: air traffic control*

There are fields where operators' analytic capacities are continuously requested and where cognitive load can quickly exert significant stress when stimuli become too burdensome. Air traffic controllers follow information coming from several different devices at the same time, and they must quickly react when directions are applied incorrectly or when an unforeseen situation arises. They encounter large volumes of complex information. Although their work environment has been adapted to present data in the clearest manner possible, situations where controllers find themselves in difficulty can still occur. In order to avoid the possibility of dangerous conditions, passive BCIs can help to adapt information interfaces according to air traffic controllers' cognitive load, thereby modulating the difficulty of the task at hand [ABB 14]. Drone controllers face similar difficulties, and in this case too, measuring cognitive load in real time can increase reliability, for example by adjusting the number of drones that an operator can handle at a given time [AFE 14].

5.3. Mental fatigue and vigilance

5.3.1. *Definition*

As reported by Oken *et al.* [OKE 06], *vigilance* is defined differently depending on the field of study and the author. Psychologists and some cognitive scientists thus use the term to designate the capacity to sustain one's attention during the realization of a task for a given period of time, and commonly talk about "vigilance decrement" when subjects' performance decreases with time spent on a task ("time-on-task"; increase in reaction time and detection accuracy). Some authors use terms other than "vigilance", but use the same operational definition. One therefore often encounters the notion of *mental fatigue*, which refers to a state that occurs when a long and tiring task that requires subjects to remain focused is performed, for example when driving a car [LAL 02, KAT 09, ZHA 12, BOR 12]. This mental fatigue affects subjects' capacity to maintain an adequate level of sustained attention and therefore affects their vigilance [BOK 05]. Finally, for neurophysiologists, this term corresponds to a level of physiological alertness on a sleep–wake continuum, without reference to a cognitive or behavioral state. In this book, we will use the term *mental fatigue* to refer to the gradual and cumulative process related to a general reduction in vigilance following an increase in *time-on-task* [LAL 02].

5.3.2. *Behavioral markers*

The state of mental fatigue is traditionally defined as a drop in behavioral performance. Subjects' reaction time thus increases, almost linearly, with an increase in time-on-task [MAC 68, GAL 77, SCH 09, KAT 09, ZHA 12, BOK 05, PAU 97]. Moreover, response accuracy drops as well [PAU 97].

5.3.3. *EEG markers*

In terms of frequencies, a drop in vigilance capacity results in a progressive increase in low-frequency EEG activity [KLI 99, OKE 06,

ZHA 12], especially in the lowest alpha frequency band [GAL 77, BOK 05] as well as in the theta band [PAU 97, BOK 05, ZHA 12]. This increase is accompanied by a decrease in high-frequency activity [KLI 99, LAL 02, OKE 06, ZHA 12, FAB 12].

With respect to evoked EEG activity, the amplitude of the P3 component decreases with a drop in vigilance [KOE 92, MUR 05, OKE 06, SCH 09]. Other authors argue that its latency also increases [KAT 09, ZHA 12]. Finally, with respect to the earliest components, the amplitude of the N1 component decreases with time-on-task at the parietal electrodes [BOK 05, FAB 12], whereas the amplitude of the N2b component increases at the central electrodes [BOK 05].

5.3.4. *Application example: driving*

Driving a car is one security and public health hazard for which passive BCIs could help to reduce the number of accidents. If a large number of simpler devices, such as eye movement recording, make it possible to detect microsleep states, EEG can help to prevent dangerous behavioral mistakes even earlier [BLA 10]. Still at the prototype stage, passive BCIs are studied using driving simulators [LAL 02]. Their performance makes it possible to hope that one day they could be used to increase road safety.

5.4. Attention

5.4.1. *Definition*

Selective temporal and spatial attention is the capacity to detect and select a specific target item that is relevant for a given task (for the purpose of deep processing) and to ignore distractors when these items are presented simultaneously or sequentially [HIL 73, POS 80]. This capacity for selective attention to important events that require a specific action can be essential in an ecological context (i.e. in day to day tasks). It is, for example, required in surveillance tasks for operators in a nuclear plant, for air traffic controllers [FU 07] or for car driving and airplane or spaceship piloting, which are more or less demanding and prolonged.

5.4.2. *Behavioral markers*

In the same way that behavioral performance is degraded with an increase in cognitive load or mental fatigue, a decrease in the attention devoted to a task or a stimulus produces a decrease in the individual's performance [POS 80, MAN 95]. Posner *et al.* thus show that we are much quicker in detection tasks when we previously pay attention to the place where the target will appear [POS 80].

5.4.3. *EEG markers*

A task typically used to study selective attention is the oddball task. This task consists of detecting a target item, which is generally rare (present in 10–20% of cases), from among a sequence of frequent distractor items (80–90% of cases) [FIT 81, FRI 01, KOK 01]. Depending on the study, subjects must count the rare target items or perform a specific response to them (e.g. press a button only for the targets, or press different buttons according to the items). In this kind of task, selective attention is accompanied by the item's occurrence probability effect. The typically observed result for this task, when performed in the visual modality, is a modulation of the P300 component evoked at parietocentral and occipital sites by processing items according to their frequency of appearance. Indeed the P300 of a target item will be more ample and longer than that of distractor item, with the amplitude difference being greater when the target item is more rare [HIL 73, PIC 92, KOL 97, FRI 01, GOM 06, FU 07]. This is why that component is also frequently studied in the context of active BCIs, more specifically for the P300 speller paradigm [GRA 10].

Other earlier neuronal components are also modified by this selective attention to a kind of items, with for example greater amplitude of the N1 component for target items than for distractor items at the vertex in the auditory modality [HIL 73], and at the parietal electrodes in the visual modality [FAB 12]. Finally, the N2b component has a larger amplitude for targets than for distractors at the central electrodes [BOK 05].

With respect to frequency EEG markers, frequent and rare item processing differs in terms of the activity evoked in the delta, theta and alpha bands at the frontocentral and parietocentral sites, with a prolongation of this activity in the case of rare items [KOL 97, BAS 92, YOR 98, ONI 09].

5.4.4. *Application example: teaching*

If there is one environment where attention tends to drastically drop despite the best intentions, it is teaching. What teacher has not feared the terrible 2:00 PM slump, right after lunch, when only empty stares respond to even the most eloquent presentations? And there is a population of students that suffers more than the others from this difficulty in selecting and processing information for prolonged periods of time; it is children suffering from attention deficit disorder. Bring the two together, add a spoonful of virtual reality, sprinkle in some passive BCIs, and you get a system that allows students to overcome their difficulties by controlling the presence of distractors in their simulated environment and rewards those that manage to remain attentive [CHO 02].

5.5. Error detection

5.5.1. *Definition*

It is the individual that is ultimately measured by the yardstick of passive BCIs. So we take the notion of an "error" in a subjective sense here. For a user, an error occurs when the result of an action is different from what he or she expected. It is not a question of whether it is "true" or "false", like evaluating responses to a questionnaire or performing a task. It is the representations that a person has of his or her environment that count. We can distinguish between the following four different types of errors [FER 08]:

– a response error is detected in operators that realize they have committed a mistake;

– an interaction error occurs when a system reacts in an unexpected manner;

– an observation error is generated by monitoring a third person;

– a *feedback* error is detected when a sanction (reward or punishment) is different from what was expected.

5.5.2. *Behavioral markers*

Although the *production* of errors is not the focus of this text, a person's capacity to detect an incongruent event directly depends on his or her vigilance state and on his or her attention level. Once an error has been detected, either the operator seeks to rectify it if he or she can influence the system, or, as is often the case in the context of learning, he or she modifies his or her internal representations to adapt to it.

5.5.3. *EEG markers*

Error detection employs evoked potentials, referred to as "error potentials" in that context, with their main recording site at FCz [SCH 00, PAR 03]. Both negative and positive potential components are studied. They make it possible not only to determine if a mistake has been detected but also to identify the kind of mistake at hand. It is even sometimes possible to measure error potentials without the person becoming aware of their mistake [NIE 01].

In the execution of a simple task like target selection, a negative component occurring after 80 ms is characteristic of a response error [FAL 00], a 250 ms delay is associated with a feedback error [HOL 02] and the late presence of a positive component denotes an interaction error [FER 08].

5.5.4. *Application example: tactile and robotic interfaces*

What can be more annoying than a computer that does not do what we ask of it? We press a button and the wrong menu appears; we press a key and it erases in a single blow the last 6 months' work – which we had never taken the time to save. There is much to be impatient about. And the cherry on top: the more interfaces evolve, the more we have the impression that computers take a sick pleasure in only doing things their way. But no need to panic; the outlook is not quite that somber. Using EEG to detect interaction errors is a step forward in creating systems that self-correct when they provide an inadequate response, for example with tactile [VI 12] or gesture [CHA 10] interfaces. This may be much more useful when what is on the other end of the wire is not a static computer but rather a robot happily bouncing around its hundreds of pounds of alloy, which is a bit more concerning. In this case too, error recognition could make it possible to correct an inadequate behavior in a matter of seconds, whether it is in interaction with a human [FER 08] or when a human is merely observing the machine [ITU 10].

5.6. Emotions

5.6.1. *Definition*

As shown by Paul Ekman through his studies of different cultures, there is a universal component in the recognition of emotions and by extension in emotion production [EKM 99]. There are different nomenclatures for classifying and describing emotions. The most commonly used distinguishes between two dimensions: *valence* and *arousal*. Valence corresponds to the "quality" of an emotion and varies from negative (e.g. "pain") to positive (e.g. "joy"). The second axis, arousal, accounts for the emotion's "intensity level". We can say that "boredom" has a low level of arousal, whereas "terror" possesses a high level of arousal [PIC 95].

5.6.2. *Behavioral markers*

We have abandoned some philosophical notions that, by opposing mind and body, tended to dissociate reason from emotions. Recent work in neuroscience has shown that the emotional component is inextricably related to decision-making mechanisms [DAM 94]. Emotions make it possible to decide more quickly between competing alternatives, with a positive valence being able to guide a choice. They also have an effect on behavior that can be described by the notions of approach and avoidance. When we compare our reaction time between negative and positive stimulation, we are quicker to avoid dangerous situations and faster to approach objects that seem to provide a reward [CHE 99].

A high arousal state brings about several physiological changes, but the inverse is also true, and an increase in metabolic activity, like heartbeat acceleration, can instill emotions and influence our reactions [SCH 62]. Studying emotions also makes it possible to uncover the differences that mark some populations. Children thus have difficulties distinguishing variations in emotional intensity [POS 05].

5.6.3. *EEG markers*

More often than with other mental states, emotional state detection using data exclusively drawn from neuroimaging lends itself to controversy. Indeed, with passive BCIs, EEG is often used, a tool that *also* records the facial muscle contractions that we have trouble avoiding making when we are experiencing heightened emotions. This accounts for the special care that must be taken when performing measurements [HEI 09].

This being said, several markers have been associated with emotional state measurements. Power spectral density in the alpha band in the frontal lobes can correlate with emotional valence. Negative valence is correlated with decreasing power in the left lobe, with a positive valence instead being related to an asymmetry within the right lobe [MOL 09]. The arousal level associated with a stimulus is itself more easily detectable in the theta band, or by studying the amplitude of the evoked potentials [MOL 09].

Beyond the signal's nature, studying the structures activated during an emotional response makes it possible to identify the trigger, either images or sounds [MUH 11].

5.6.4. *Application example: communication and personal development*

No one will be surprised to read that emotions make it possible to directly communicate one's feelings. That seemingly banal statement, nevertheless, indicates the potential for BCIs that can be used to reveal the affection we feel toward lifeless objects – such as robots – be it inclination or aversion [STR 14]. Fortunately, new technologies also help to increase the social link between flesh and bone humans. Even though in this case the measurement relies more heavily on muscle activity than on brain waves, Necomimi ears developed by the *neurowear* team are a shocking example of the kind of enhancement we could benefit from. They are large plush cat ears that are supposed to move in such a way that reflects the user's emotional state. Interacting with peers becomes a much more honest affair. For those who already have trouble tolerating those around them, they may rest assured that EEG measures can also help them take charge, like in social stress situations [JEU 14]. This kind of research is part of the trend to employ *neurofeedback*, as will be discussed in Chapter 13.

5.7. Conclusions

Thanks to technological developments brought about by active BCIs, passive BCIs have been able to see the light of day and are fully expanding. Their contributions are numerous, going from basic research in cognitive neuroscience (with a better understanding of the neurophysiological phenomena that underlie cognitive functions) to human factors (with the development of embedded systems that make it possible to enhance safety in high-risk work environments or piloting situations), as well as the improvement of BCIs. Although several markers of different mental states have already been identified, they often intersect, which can thus disturb recognition systems in ecological settings. Moreover, interactions between mental states can alter markers' relevance. Research should therefore focus on identifying markers that are robust to the type of task performed by operators and to interactions between mental states.

5.8. Bibliography

[ABB 14] ABBASS H.A., TANG J., AMIN R. *et al.*, "Augmented cognition using real-time EEG-based adaptive strategies for air traffic control", *HFES'14 Human Factors and Ergonomics Society 2014 Annual Meeting*, vol. 58, no. 1, pp. 230–234, 2014.

[AFE 14] AFERGAN D., PECK E.M., SOLOVEY E.T. *et al.*, "Dynamic difficulty using brain metrics of workload", *CHI'14*, ACM Press, New York, NY, USA, pp. 3797–3806, 2014.

[ALL 08] ALLISON B.Z., POLICH J., "Workload assessment of computer gaming using a single-stimulus event-related potential paradigm", *Biological Psychology*, vol. 77, no. 3, pp. 277–283, 2008.

[ANT 10] ANTONENKO P., PAAS F., GRABNER R. *et al.*, "Using electroencephalography to measure cognitive load", *Educational Psychology Review*, vol. 22, pp. 425–438, 2010.

[BAS 92] BASAR-EROGLU C., BASAR E., DEMIRALP T. *et al.*, "P300-response: possible psychophysiological correlates in delta and theta frequency channels. A review", *International Journal of Psychophysiology*, vol. 13, no. 2, pp. 161–179, 1992.

[BER 07] BERKA C., LEVENDOWSKI D.J., LUMICAO M.N. *et al.*, "EEG correlates of task engagement and mental workload in vigilance, learning, and memory tasks", *Aviation, Space, and Environmental Medicine*, vol. 78, no. 5 Suppl, pp. B231–B244, 2007.

[BLA 10] BLANKERTZ B., TANGERMANN M., VIDAURRE C. *et al.*, "The Berlin brain–computer interface: non-medical uses of BCI technology.", *Frontiers in Neuroscience*, vol. 4, p.198, 2010.

[BOK 05] BOKSEM M., MEIJMAN T., LORIST M., "Effects of mental fatigue on attention: an ERP study", *Cognitive Brain Research*, vol. 25, no. 1, pp. 107–116, 2005.

[BOR 12] BORGHINI G., ASTOLFI L., VECCHIATO G. *et al.*, "Measuring neurophysiological signals in aircraft pilots and car drivers for the assessment of mental workload, fatigue and drowsiness", *Neuroscience & Biobehavioral Reviews*, vol. 44, pp. 58–75, 2012.

[CAI 07] CAIN B., A Review of the Mental Workload Literature 1.0, Technical Report, Defence Research and Development Canada Toronto, 2007.

[CHA 10] CHAVARRIAGA R., BIASIUCCI A., FORSTER K. *et al.*, "Adaptation of hybrid human-computer interaction systems using EEG error-related potentials", *IEEE Engineering in Medicine and Biology Society*, vol. 2010, pp. 4226–4229, 2010.

[CHE 99] CHEN M., BARGH J.A., "Consequences of automatic evaluation: immediate behavioral predispositions to approach or avoid the stimulus", *Personality and Social Psychology Bulletin*, vol. 25, no. 2, pp. 215–224, 1999.

[CHO 02] CHO B., LEE J., KU J. *et al.*, "Attention enhancement system using virtual reality and EEG biofeedback", *Proceedings of IEEE Virtual Reality*, vol. 2002, pp. 156–163, 2002.

[DAM 94] DAMASIO A.R., *Descartes' Error: Emotion, Reason, and the Human Brain*, G.P. Putnam's Sons, New York, 1994.

[EKM 99] EKMAN P., "Facial expressions", in DALGLEISH T., POWER M.J. (eds), *Handbook of Cognition and Emotion*, John Wiley & Sons Ltd, Chichester, England, pp. 301–320, 1999.

[FAB 12] FABER L.G., MAURITS N.M., LORIST M.M., "Mental fatigue affects visual selective attention", *PLoS ONE*, vol. 7, no. 10, p. e48073, 2012.

[FAL 00] FALKENSTEIN M., HOORMANN J., CHRIST S. *et al.*, "ERP components on reaction errors and their functional significance: a tutorial", *Biological Psychology*, vol. 51, nos. 2–3, pp. 87–107, 2000.

[FER 08] FERREZ P.W., MILLAN J.D.R., "Error-related EEG potentials generated during simulated brain–computer interaction", *IEEE Transactions in Biomedical Engineering*, vol. 55, no. 3, pp. 923–929, 2008.

[FIT 81] FITZGERALD P.G., PICTON T.W., "Temporal and sequential probability in evoked potential studies", *Canadian Journal of Psychology*, vol. 35, no. 2, pp. 188–200, 1981.

[FRI 01] FRIEDMAN D., CYCOWICZ Y.M., GAETA H., "The novelty P3: an event-related brain potential (ERP) sign of the brain's evaluation of novelty", *Neuroscience Biobehavioral Review*, vol. 25, no. 4, pp. 355–373, 2001.

[FU 07] FU S., PARASURAMAN R., "Event-related potentials (ERPs) in neuroergonomics", in PARASURAMAN R., RIZZO M., (eds), *Neuroergonomics: The Brain at Work*, Oxford University Press, Inc., New York, NY, pp. 15–31, 2007.

[GAL 77] GALE A., DAVIES R., SMALLBONE A., "EEG correlates of signal rate, time in task and individual differences in reaction time during a five-stage sustained attention task", *Ergonomics*, vol. 20, no. 4, pp. 363–376, 1977.

[GEO 10] GEORGE L., LÉCUYER A., "An overview of research on "passive" brain–computer interfaces for implicit human-computer interaction", *ICABB '10*, 2010.

[GEV 00] GEVINS A., SMITH M.E., "Neurophysiological measures of working memory and individual differences in cognitive ability and cognitive style", *Cerebral Cortex*, vol. 10, no. 9, pp. 829–839, 2000.

[GEV 03] GEVINS A., SMITH M.E., "Neurophysiological measures of cognitive workload during human-computer interaction", *Theoretical Issues in Ergonomics Science*, vol. 4, nos. 1–2, pp. 113–131, 2003.

[GEV 07] GEVINS A., SMITH M.E., "Electroencephalography (EEG) in neuroergonomics", in PARASURAMAN R., RIZZO M. (eds), *Neuroergonomics: The Brain at Work*, Oxford University Press, Inc., New York, NY, pp. 15–31, 2007.

[GOM 06] GOMARUS H.K., ALTHAUS M., WIJERS A.A. *et al.*, "The effects of memory load and stimulus relevance on the EEG during a visual selective memory search task: an ERP and ERD/ERS study", *Clinical Neurophy*chology, vol. 117, no. 4, pp. 871–884, 2006.

[GRA 10] GRAINMANN B., ALLISON B., PFURTSCHELLER G., "Brain–computer interfaces: a gentle introduction", *Brain–Computer Interfaces: Revolutionizing Human-computer Interaction*, Springer, Berlin, Heidelberg, pp. 1–25, 2010.

[HEI 09] HEINGARTNER D., "Mental block", *IEEE Spectrum*, vol. 46, no. 1, pp. 42–43, 2009.

[HIL 73] HILLYARD S.A., HINK R.F., SCHWENT V.L. *et al.*, "Electrical signs of selective attention in the human brain", *Science*, vol. 182, no. 4108, pp. 177–180, 1973.

[HOL 02] HOLROYD C.B., COLES M.G., "The neural basis of human error processing: reinforcement learning, dopamine, and the error-related negativity", *Psychological Review*, vol. 109, no. 4, pp. 679–709, 2002.

[HOL 09] HOLM A., LUKANDER K., KORPELA J. *et al.*, "Estimating brain load from the EEG", *Scientific World Journal*, vol. 9, pp. 639–651, 2009.

[ITU 10] ITURRATE I., MONTESANO L., MINGUEZ J., "Single trial recognition of error-related potentials during observation of robot operation", *IEEE Engineering in Medicine and Biology Society*, vol. 2010, pp. 4181–4184, 2010.

[JEU 14] JEUNET C., MÜHL C., LOTTE F., "Design and validation of a mental and social stress induction protocol towards load-invariant physiology-based detection", *PhyCS'14*, 2014.

[KAT 09] KATO Y., ENDO H., KIZUKA T., "Mental fatigue and impaired response processes: event-related brain potentials in a Go/NoGo task", *International Journal of Psychophysiol*ogy, vol. 72, no. 2, pp. 204–211, 2009.

[KLI 99] KLIMESCH W., "EEG alpha and theta oscillations reflect cognitive and memory performance: a review and analysis", *Brain Research Re*view, vol. 29, nos. 2–3, pp. 169–195, 1999.

[KOE 92] KOELEGA H.S., VERBATEN M.N., VAN LEEUWEN T.H. *et al.*, "Time effects on event-related brain potentials and vigilance performance", *Biological Psychol*ogy, vol. 34, no. 1, pp. 59–86, 1992.

[KOK 01] KOK A., "On the utility of P3 amplitude as a measure of processing capacity", *Psychophysiology*, vol. 38, no. 3, pp. 557–577, 2001.

[KOL 97] KOLEV V., DEMIRALP T., YORDANOVA J. *et al.*, "Time-frequency analysis reveals multiple functional components during oddball P300", *Neuroreport*, vol. 8, no. 8, pp. 2061–2065, 1997.

[LAL 02] LAL S.K., CRAIG A., "Driver fatigue: electroencephalography and psychological assessment", *Psychophysiol*ogy, vol. 39, no. 3, pp. 313–321, 2002.

[MAC 68] MACKWORTH J.F., "vigilance, arousal, and habituation", *Psychological Review*, vol. 75, no. 4, pp. 308–322, 1968.

[MAN 95] MANGUN G.R., "Neural mechanisms of visual selective attention", *Psychophysiology*, vol. 32, no. 1, pp. 4–18, 1995.

[MIL 11] MILLER M.W., RIETSCHEL J.C., MCDONALD C.G. *et al.*, "A novel approach to the physiological measurement of mental workload", *International Journal of Psychophysiology*, vol. 80, no. 1, pp. 75–78, 2011.

[MIS 06] MISSONNIER P., DEIBER M.-P., GOLD G. *et al.*, "Frontal theta event-related synchronization: comparison of directed attention and working memory load effects", *Journal of Neural Trans*port, vol. 113, no. 10, pp. 1477–1486, 2006.

[MOL 09] MOLINA G.G., TSONEVA T., NIJHOLT A., "Emotional brain–computer interfaces", *IJAACS*, vol. 6, no. 1, p. 9, 2009.

[MUH 11] MUHL C., BROUWER A., VAN WOUWE N. *et al.*, "Modality-specific affective responses and their Implications for affective BCI", *Fifth International Brain–Computer Interface Conference*, pp. 120–123, 2011.

[MUR 05] MURATA A., UETAKE A., TAKASAWA Y., "Evaluation of mental fatigue using feature parameter extracted from event-related potential", *International Journal of Industrial Ergon*omics, vol. 35, no. 8, pp. 761–770, 2005.

[NAT 81] NATANI K., GOMER F.E., "Electrocortical activity and operator workload: a comparison of changes in the electroencephalogram and in event-related potentials", *McDonnell Douglas Technical Report*, pp. 188–200, 1981.

[NIE 01] NIEUWENHUIS S., RIDDERINKHOF K.R., BLOM J. *et al.*, "Error-related brain potentials are differentially related to awareness of response errors: evidence from an antisaccade task", *Psychophysiol*ogy, vol. 38, no. 5, pp. 752–760, 2001.

[OKE 06] OKEN B., SALINSKY M., ELSAS S., "Vigilance, alertness, or sustained attention: physiological basis and measurement", *Clinical Neurophys*ics, vol. 117, no. 9, pp. 1885–1901, 2006.

[ONI 09] ONIZ A., BAAR E., "Prolongation of alpha oscillations in auditory oddball paradigm", *International Journal of Psychophysiol*ogy, vol. 71, no. 3, pp. 235–241, 2009.

[OSS 11] OSSANDON T., JERBI K., VIDAL J.R. *et al.*, "Transient suppression of broadband gamma power in the default-mode network is correlated with task complexity and subject performance", *Journal of Neurosci*ence, vol. 31, no. 41, pp. 14521–14530, 2011.

[PAR 03] PARRA L.C., SPENCE C.D., GERSON A.D. *et al.*, "Response error correction—a demonstration of improved human-machine performance using real-time EEG monitoring.", *IEEE Engineering in Medicine and Biology Society*, vol. 11, no. 2, pp. 173–177, 2003.

[PAU 97] PAUS T., ZATORRE R.J., HOFLE N. *et al.*, "Time-related changes in neural systems underlying attention and arousal during the performance of an auditory vigilance task", *Journal of Cognitive Neurosci*ence, vol. 9, no. 3, pp. 392–408, 1997.

[PIC 92] PICTON T.W., "The P300 wave of the human event-related potential", *Journal of Clinical Neurophysiology*, vol. 9, no. 4, pp. 456–479, 1992.

[PIC 95] PICARD R.W., Affective computing, Report no. 321, MIT Media Laboratory, 1995.

[POS 80] POSNER M., "Orienting of attention", *Quaterly Journal of Experimental Psychology*, no. 32, pp. 3–25, 1980.

[POS 05] POSNER J., RUSSELL J.A., PETERSON B.S., "The circumplex model of affect: an integrative approach to affective neuroscience, cognitive development, and psychopathology", *Developmental Psychopathology*, vol. 17, no. 3, pp. 715–34, 2005.

[PUT 10] PUTZE F., JARVIS J., SCHULTZ T., "Multimodal recognition of cognitive workload for multitasking in the car", *ICPR'10*, pp. 3748–3751, 2010.

[SCH 62] SCHACHTER S., SINGER J., "Cognitive, social, and physiological determinants of emotional state", *Psychological Review*, vol. 69, no. 5, pp. 379–399, 1962.

[SCH 00] SCHALK G., WOLPAW J.R., MCFARLAND D.J. *et al.*, "EEG-based communication: presence of an error potential", *Clinical Neurophysiology*, vol. 111, no. 12, pp. 2138–2144, 2000.

[SCH 04] SCHULTHEIS H., JAMESON A., "Assessing cognitive load in adaptive hypermedia systems: physiological and behavioral methods", in HUTCHISON D., KANADE T., KITTLER J. *et al.*, (eds), *Adaptive Hypermedia and Adaptive Web-Based Systems*, vol. 3137, Springer, Berlin, Heidelberg, 2004.

[SCH 09] SCHMIDT E., SCHRAUF M., SIMON M. *et al.*, "Drivers' misjudgement of vigilance state during prolonged monotonous daytime driving", *Accident Analysis & Prevention*, vol. 41, no. 5, pp. 1087–1093, 2009.

[STE 66] STERNBERG S., "High-speed scanning in human memory", *Science*, vol. 153, no. 3736, pp. 652–654, 1966.

[STE 69] STERNBERG S., "Memory-scanning: mental processes revealed by reaction-time experiments", *American Scientist*, vol. 57, no. 4, pp. 421–457, 1969.

[STR 14] STRAIT M., SCHEUTZ M., "Using near infrared spectroscopy to index temporal changes in affect in realistic human-robot interactions", *PhyCS '14*, pp. 385–392, 2014.

[VI 12] VI C., SUBRAMANIAN S., "Detecting error-related negativity for interaction design", *CHI'12*, ACM Press, New York, NY, USA, pp. 493–502, 2012.

[YOR 98] YORDANOVA J., KOLEV V., "Single-sweep analysis of the theta frequency band during an auditory oddball task", *Psychophysiology*, vol. 35, no. 1, pp. 116–126, 1998.

[ZAN 11] ZANDER T.O., KOTHE C., "Towards passive brain–computer interfaces: applying brain–computer interface technology to human-machine systems in general", *Journal of Neural Engineering*, vol. 8, no. 2, 2011.

[ZHA 12] ZHAO C., ZHAO M., LIU J. *et al.*, "Electroencephalogram and electrocardiograph assessment of mental fatigue in a driving simulator", *Accident Analysis & Prevention*, vol. 45, pp. 83–90, 2012.

PART 2

Signal Processing and Machine Learning

6

Electroencephalography Data Preprocessing

6.1. Introduction

Electroencephalography[1] provides measurements of electrical potential in the form of a time signal for each electrode. Although they only partially reflect the brain's underlying electrophysiological phenomena (see Chapter 3), these measurements contain a significant amount of information, which is used in clinical diagnosis and cognitive sciences, as well brain computer interfaces. The information provided by EEG is varied in nature: it can show variations of electric potential in amplitude at certain times and at certain frequencies, or variations in its spatial distribution. Accordingly, to observe these phenomena, a variety of methods are used in order to analyze signals in time, in frequency or in space. Finally, statistical analysis is also necessary to assess the stability of the information extracted with respect to different parameters (in time, through subjects, through different populations, between different experimental conditions).

Before being interpreted, an EEG should be properly preprocessed in order to denoise or filter the measurements and to extract different

---

Chapter written by Maureen CLERC.

1 This whole chapter, which deals with electroencephalography, also applies to magnetoencephalography described in Chapter 2.

components of interest to the study at hand. In this chapter, we will see that the data can be represented from a variety of viewpoints (temporal, frequency, time–frequency, spatial, statistical), offering many possibilities for preprocessing.

Section 6.2 focuses on EEG acquisition, which in practice reduces the amount of available information. The subsequent sections discuss different representations in time and frequency in sections 6.3 and 6.4 respectively, and then, in section 6.5, combined time–frequency representations, which make it possible to analyze non-stationary signals. Section 6.6 discusses a variety of spatial representations, from topographic representations on the scalp's surface, to the reconstruction of electrical brain sources. Finally, section 6.7 focuses on statistical analyses aimed at extracting or separating components.

Some methods presented in this chapter have also been addressed in Chapter 2 on brain imaging, of which they are an integral part (time–frequency representations, source localization, independent component analysis (ICA)).

6.2. Principles of EEG acquisition

Because of its very low amplitude, which is in the order of 10–100 μV, EEG data can only be measured through an amplifier. This amplifier provides one time-dependent signal per measured channel. Originally, EEG amplifiers were analog, using a pen to trace signals on a roll of paper, much like a seismograph. Nowadays, most amplifiers are digital: they perform an analog–digital conversion and supply data in the form of sampled and quantized signals.

6.2.1. *Montage*

An EEG carries out measurements through a set of electrodes placed on the scalp, as explained in Chapter 1. Data come in the form of time signals, each corresponding to an acquisition channel. However, a channel's signal does not directly correspond to an electrode's electrical potential, since only *electrical potential differences* are measurable. There are different ways, called *montages*, of measuring these potential differences. The most common types are monopolar and bipolar.

In a monopolar montage, the same reference is used for each channel. This reference can be derived from a particular electrode (called the *reference electrode*), or from the average electric potential over all electrodes (called the *average reference*).

The bipolar montage measures differences in potential between neighboring electrodes. Each channel therefore has a different reference. A bipolar channel is especially sensitive to the electrical activity coming from sources located between the two electrodes.

The choice of montage used to be crucial in the times when analog amplifiers traced measurements on strips of paper. But with digital amplifiers, it is now possible to process the data *a posteriori* in order to transform one montage to another.

6.2.2. *Sampling and quantification*

With digital amplifiers, EEG recordings are not continuous but discrete. Specifically, the data are sampled, i.e. measured at instants separated by the *sampling interval*. This is usually the same throughout the duration of the recording. The *sampling rate* (or sampling frequency) is the inverse of the sampling interval. For example, a signal sampled at 100 Hz has a sampling interval of 10 ms.

The sampling frequency (or interval) determines the temporal resolution of observable phenomena, as well as the extent of the analyzable frequency spectrum (see section 6.4.1). Shannon's sampling theory calls for a sampling frequency greater than twice the target maximum frequency [MAL 08].

The analog–digital conversion performed by EEG amplifiers quantifies the signals: their amplitude is coded with a finite number of bits. For instance, 16 bits provide 2^{16} values to encode the whole dynamic range of the EEG. With 16 bits, values between $\pm 600\ \mu V$ have an amplitude precision of $0.0183\ \mu V$.

6.3. Temporal representation and segmentation

The most basic representation of EEG data provided directly by the amplifier is temporal. Raw EEG data comes in the form of time *signals*, each corresponding to a channel. Observing these signals during their acquisition is important in order to ascertain that they are of good quality.

6.3.1. *Segmentation*

The signals measured by the amplifier can be accompanied by additional time stamps known as *events*. Each event is associated with a date (a time sample) and a label specifying its nature. Events are sometimes marked by the experimenter using a specific software, during data acquisition or review. In brain–computer interfaces, it is more common for marking to be performed automatically during acquisition. The technical details of this marking process are explained in Chapter 7 of Volume 2 [CLE 16]

Events provide temporal information that makes it possible to define segments of particular interest for data analysis through a time window (these segments are also referred to as "epochs"). This procedure is called data *segmentation* or *epoching* . In an EEG data set, we thus have *epochs* or *trials* of similar events. With these trials (also called realizations in statistical terms), and machine learning methods, it will be possible to automatically classify new data whose nature is unknown *a priori*.Trials are also important for EEG analysis, since they make it possible to obtain enough data to reduce the effect of noise. Averaging the repetitions together yields "evoked potentials" (also see Chapter 2).

6.3.2. *Time domain preprocessing*

Noise reduction: Some parts of the signal can be eliminated simply because they display undesirable behavior (e.g. because of displaced electrodes that have lost contact with the scalp and only measure noise).

Component separation: The signal components of interest can be identified directly in temporal signals, especially in low-frequency components (such as the P300), or in large amplitude changes (such as slow cortical potentials, or the spikes observed in epilepsy). However, the activity measured on the sensors results from a combination of several brain sources, and as we shall see in section 6.6.2, these temporal components are generally better identified after spatial filtering.

6.4. Frequency representation

Since the EEG recordings in humans were carried out by Hans Berger, its oscillatory character has been evident. We have seen in previous chapters that EEG contains oscillations (alpha and beta waves, etc.) from different brain regions. To represent the EEG spectrum, that is to say its frequency variations, it is necessary to use the Fourier transform [BRI 01].

6.4.1. *Fourier transform*

We define a function's (continuous) Fourier transform as

$$\hat{f}(\omega) = \int_{-\infty}^{\infty} f(t) \exp(-i\omega\, t) dt$$

Consider a signal f measured on the interval $[0, T]$ with N discrete samples separated by τ (the sampling interval): $f[n] \doteq f(n\tau)$, $n = 0 \ldots N-1$. The discrete Fourier transform is defined for $k = 0$ to $N-1$ by

$$\hat{f}[k] = \sum_{n=0}^{N-1} f[n] \exp^{-2i\pi kn/N}$$

The discrete Fourier transform of f is a sequence of complex numbers $\hat{f}[k] = c_k \exp(i\phi_k)$, where c_k (respectively ϕ_k) represents the amplitude (respectively phase) spectrum.

The definition of the discrete Fourier transform involves only N samples of the signal, and due to this finite length, it is mathematically defined as the Fourier series of an N-periodic Dirac comb obtained by *periodization* of $f[n], n = 0, \ldots, N-1$. This is of considerable importance if we consider regularity properties, because periodization can artificially introduce discontinuities. The Fourier transform of a discontinuous function decreases slowly at high frequencies, causing artificial components to appear in its spectrum. This frequency alteration is called the Gibbs phenomenon.

It is therefore preferable to apply the Fourier transform to signals whose periodization is not discontinuous: before applying the Fourier transform, it is often recommended to multiply the signal by a window function, equal to zero at both ends $t = 0$ and $t = T$. This window (called a *tapering* window) has the effect of blurring the periodization discontinuity without dramatically altering the original signal's Fourier transform. Many window functions can accomplish this (Gaussian, Hamming, Blackman, Hanning, etc.), and it is necessary to compare each one's relative advantage in terms of frequency resolution, frequency spreading and amplitude modifications.

A signal $f[n] = f(n\tau)$, $n = 0 \ldots N-1$ sampled at frequency $F = 1/\tau$ (the inverse of the sampling interval τ), has a discrete Fourier transform that includes frequencies between $-F/2$ and $F/2$ with a frequency resolution (the distance between two samples) of $F/N \approx \frac{1}{N\tau} = \frac{1}{T}$. It is therefore important to pay attention to both the sampling frequency and the signal length when it comes to its spectrum. To summarize:

– the sampling rate limits the extent of the analyzable Fourier spectrum, as mentioned above;

– on the other hand, the signal duration T determines the spectrum's frequency resolution, i.e. the minimum frequency distance between components such that they are distinguishable.

The most common digital method for calculating the discrete Fourier transform is the fast Fourier transform (FFT), a recursive algorithm that is valuable because of its low computational complexity: it only grows by a factor of $N \log N$ with the number of samples N. It is important to recall that when the signal length N increases, the estimation of its spectrum does not automatically become more precise. Straightforward estimators of spectral density are indeed biased, and it is necessary to resort to a periodogram to correct for this phenomenon [BRO 91, BRI 01].

6.4.2. *Frequency filtering*

Noise reduction: it is possible to eliminate artifacts that occur in specific frequency bands through filtering. A high-pass filter removes low frequencies, which is useful to correct for occasionally observed low-frequency fluctuations, especially with dry electrodes. A low-pass filter removes high frequencies, eliminating some sources of noise, such as electrical activity coming from muscles. The power line current produces an artifact on the

EEG channels at a given frequency[2], which can be eliminated by a band-stop filter (notch filter).

Component separation: many neurophysiological BCI markers are in a specific frequency band, as seen in Chapter 4, which may even be specific to each subject (for example the motor rhythm μ). However, for steady-state visual evoked potentials, the frequency of interest is not subject specific, since it is usually twice the stimulation frequency. Frequency analysis of the signal is important in order to filter signals in the appropriate band. Filtering must, however, be applied carefully because, for non-stationary signals, it may introduce artifacts. This may be the case, for example, for epileptic spikes, which, when filtered at high frequency, may suggest the presence of oscillations that in fact result only from the Gibbs phenomenon [BÉN 10].

6.5. Time–frequency representations

The Fourier transform is a powerful tool for decomposing EEG signals into oscillatory components, and thus for analyzing brain rhythms. However, it is rare for these oscillations to remain stationary for the duration of the recording. Rather, it is common to observe transient events, i.e. oscillation "bursts". For this reason, it is common practice to perform frequency analysis in a time window centered around a chosen moment, while allowing this time window to move along the time axis. This is known as a *sliding window*, and leads to a time–frequency representation indexed by two variables: the time point where the decomposition is centered and the frequency of each Fourier component.

6.5.1. *Time–frequency atom*

Representing a time–frequency signal involves analyzing it around a time–frequency position (u, ξ). A time–frequency atom is the building block that localizes the analysis in time and frequency [FLA 93, MAL 08]. It is a normalized function ϕ ($\|\phi\|^2 = \int_{-\infty}^{\infty} |\phi(t)|^2 \, dt = 1$), centered in time around $u = \int_{-\infty}^{\infty} t \, |\phi(t)|^2 \, dt$, and in frequency around $\xi = \frac{1}{2\pi} \int_{-\infty}^{\infty} \omega \, |\hat{\phi}(\omega)|^2 \, d\omega$.

2 50 Hz in Europe, Asia and Africa; 60 Hz in North America.

Analyzing a signal f around (u, ξ) amounts to calculating the inner product $\langle f, \phi \rangle$ of f and ϕ (which involves the complex conjugate of ϕ, denoted ϕ^*):

$$\langle f, \phi \rangle = \int_{-\infty}^{\infty} f(t)\, \phi^*(t)\, dt$$

Suppose that the function ϕ is localized around a position u (i.e. it vanishes outside a neighborhood of u). Then, $\langle f, \phi \rangle$ depends only on the values taken by $f(t)$ in the neighborhood of $t = u$.

Similarly, in terms of frequency, using Parseval's formula, the inner product can be rewritten as:

$$\langle f, \phi \rangle = \frac{1}{2\pi} \int_{-\infty}^{\infty} \hat{f}(\omega)\, \hat{\phi}^*(\omega)\, d\omega$$

If the function $\hat{\phi}$ is localized around a frequency ξ, then $\langle f, \phi \rangle$ depends only on the values taken by $\hat{f}(\omega)$ in the neighborhood of $\omega = \xi$.

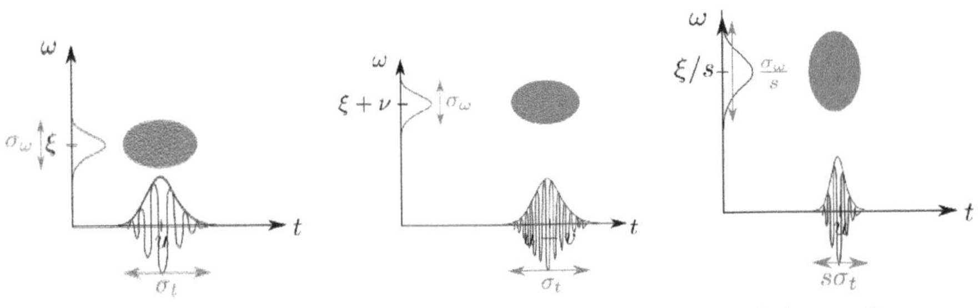

Figure 6.1. *The translation and modulation of an atom alters its time–frequency position without changing its time–frequency spread, whereas scaling alters both localization and spread*

The temporal spread of a time–frequency atom around its average position u is given by σ_t:

$$\sigma_t^2 = \int_{-\infty}^{\infty} (t - u)^2\, |\phi(t)|^2\, dt$$

The smaller the temporal spread, σ_t, the more the function is concentrated around its average position. The translation of a function changes its average position, without changing its time spread (Figure 6.1(b)).

Similarly, we can measure the frequency spread σ_ω of ϕ around ξ:

$$\sigma_\omega^2 = \frac{1}{2\pi} \int_{-\infty}^{\infty} (\omega - \xi)^2 \, |\hat{\phi}(\omega)|^2 \, d\omega \qquad [6.1]$$

Modulating a function (i.e. multiplying it by a complex exponential $e^{i\nu t}$) moves its average frequency by a factor of ν, but it does not change its frequency spread (Figure 6.1(b)).

There is a limit to joint time–frequency concentration. The more a function is concentrated in time, the less it is concentrated in frequency. This principle is expressed mathematically (under certain regularity conditions for the function ϕ) by the equation: $\sigma_t \, \sigma_\omega \geq \frac{1}{2}$. Gaussian functions are the only functions for which this relation is in fact an equality: they have a maximal time–frequency concentration.

NOTE.– In the context of quantum physics, a similar result, called "uncertainty principle", has been shown by Weyl: with increasing certainty about a particle's position comes increased uncertainty about its momentum.

6.5.2. *Short-time Fourier transform*

The short-term Fourier transform (STFT) consists of constructing a time-frequency atom around (u, ξ) by translating and modulating a symmetric real function g, supported within a time interval: $g_{u,\xi}(t) = g(t-u)e^{i\xi t}$. The function g can, for example, be a window function which is smoothed at its extremities to avoid discontinuities.

The STFT of f is obtained by computing the scalar product with the atom $g_{u,\xi}$:

$$\langle f, g_{u,\xi} \rangle = \int_{-\infty}^{\infty} f(t) g(t-u) e^{-i\xi t} dt$$

The resolution of the STFT around an instant u and a frequency ξ is uniform in the time–frequency plane. It is independent of u and ξ, but it depends on the concentration of g in time (its support) and frequency (the decay of $\hat{g}(\omega)$ at high frequencies, which is related to the regularity of g).

The *spectrogram*, energy density in time–frequency, is the squared amplitude of the STFT: $P_S f(u, \xi) = |\langle f, g_{u,\xi} \rangle|^2$.

The STFT is very useful for extracting features of brain activity for brain–computer interfaces. It is important to choose an analysis window g whose time and frequency concentrations are well adapted to the phenomena to explore.

6.5.3. *Wavelet transform*

As we have seen, the STFT has a uniform solution in time–frequency, which depends on the time–frequency concentration of g. But for certain types of signals, which have non-uniform variations at different frequencies or at different times, it is sometimes preferable to analyze the phenomena with a resolution that depends on frequency. For such signals, the wavelet transform may be an appropriate choice.

A wavelet is a normalized time–frequency atom ψ, whose average is zero $\int \psi(t)\,dt = 0$, and whose average position is also zero $\int t|\psi(t)|^2\,dt = 0$. A family of time–frequency atoms is constructed by translating ψ to position u and applying a scaling factor $s > 0$:

$$\psi_{u,s}(t) = \frac{1}{\sqrt{s}}\psi\left(\frac{t-u}{s}\right)$$

Since ψ is centered around 0, $\psi_{u,s}$ is centered around $t = u$. Recalling that the time spread σ_t of ψ is defined by $\sigma_t^2 = \int_{-\infty}^{+\infty} t^2\,|\psi(t)|^2\,dt$, a change in variable $t \mapsto \frac{t-u}{s}$ shows that the time spread of the atom $\psi_{u,s}$ is $s\sigma_t$:

$$\int_{-\infty}^{+\infty} (t-u)^2\,|\psi_{u,s}(t)|^2\,dt = s^2\,\sigma_t^2 \qquad [6.2]$$

Note η the average frequency of $\hat{\psi}$: $\eta = \frac{1}{2\pi}\int_{-\infty}^{+\infty} \omega\,|\hat{\psi}(\omega)|^2\,d\omega$. The Fourier transform $\psi_{u,s}$ corresponds to scaling $\hat{\psi}$ by a factor of $1/s$, and modulating it by u: $\hat{\psi}_{u,s}(\omega) = \sqrt{s}\,\hat{\psi}(s\omega)\,\exp(-i\omega u)$. The average frequency of ψ_{us} is therefore η/s, and its frequency spread is $\frac{\sigma_\omega}{s}$:

$$\frac{1}{2\pi}\int_0^{+\infty}\left(\omega-\frac{\eta}{s}\right)^2\left|\hat{\psi}_{u,s}(\omega)\right|^2 d\omega=\frac{\sigma_\omega^2}{s^2}$$

where σ_ω is the frequency spread of the wavelet ψ defined by equation [6.1].

Figure 6.1(c) illustrates these changes in the time–frequency spread induced by scaling.

Wavelet transform or multiscale filtering: the wavelet transform of f at time position u and scale s is defined by its inner product with $\psi_{u,s}$

$$\langle f,\psi_{u,s}\rangle=\int_{-\infty}^{+\infty} f(t)\,\frac{1}{\sqrt{s}}\,\psi^*\left(\frac{t-u}{s}\right)\,dt \qquad [6.3]$$

The wavelet transform $\langle f,\psi_{u,s}\rangle$, denoted by $W_s(u)$, amounts to filtering f with a filter scaled by s. Indeed, the scalar product [6.3] above can be seen as the convolution of f with the function $\tilde{\psi}_s(t)=\frac{1}{\sqrt{s}}\psi^*(-t/s)$. The Fourier transform of $W_s(u)$ with respect to u can thus be written as a product:

$$\widehat{W_s}(\xi)=\sqrt{s}\hat{f}(\xi)\hat{\psi}^*(s\omega)$$

This can be interpreted as the filtering of f by a high-pass filter scaled by s, since a wavelet ψ has an average of zero, so $\hat{\psi}(0)=\int_{-\infty}^{\infty}\psi(t)\,dt=0$.

The *scalogram* is the time–frequency energy of f $|\langle f,\psi_{u,s}\rangle|^2=\left|\langle f,\psi_{u,\frac{\eta}{\xi}}\rangle\right|^2$.

The time–frequency resolution of the wavelet transform and the scalogram are related to the time–frequency spread of atoms $\psi_{u,s}$. According to our previous remarks on the relationship between time–frequency spread and scale, the temporal resolution increases as the frequency increases (or when the scale s decreases). Conversely, the frequency resolution decreases as the inverse of the central frequency.

6.5.4. *Time–frequency transforms of discrete signals*

We have described the short-time Fourier transforms and wavelet transforms for continuous functions $f(t)$, but in practice they must be applied to discrete signals of the form $\{f[n], n = 1 \dots N\}$.

Time–frequency transforms of those discrete signals, whether wavelets or STFTs, may take two different forms, depending on the way the time–frequency plane is sampled.

In an *orthogonal transform*, exactly N samples are used in the time–frequency plane, i.e. the transform has the same dimensionality as the initial signal.

In a *redundant transform*, the time–frequency plane is sampled more densely, resulting in more samples than the original N samples of f.

The redundant transform may seem less interesting, because it uses more data to represent the same information. However, only a redundant transform provides the property of invariance by translation, which is very useful if there is uncertainty on the exact time when the activity of interest occurs.

Orthogonal wavelet transforms can be calculated with the fast wavelet transform, a fast algorithm with $N \log N$ complexity [MAL 08]. This type of wavelet transform is useful for signal or image compression applications (see JPEG 2000), but for EEG signal analysis, redundant wavelet transforms are preferable.

6.5.5. *Toward other redundant representations*

Time–frequency atoms are convenient for representing localized events in both time and frequency, as we have seen. When grouped together in a "dictionary" (term for a large collection of atoms providing a redundant representation of signals), the atoms most accurately representing the signals can be selected by algorithms that promote sparsity. As examples of such algorithms, we can mention Matching Pursuit [MAL 93] and its multitrial [BÉN 09] or multichannel [GRI 08] extensions.

Despite their importance, we must admit that due to their smooth, symmetric time-course, conventional time–frequency atoms (e.g. Gabor wavelets) are quite different from the neurophysiological events they seek to represent. It is often necessary to use several atoms to represent a single event, which is detrimental to interpretation. However, dictionaries can be enriched in order to contain atoms of the desired shape [SPU 15].

Since the desired shape is rarely known beforehand, the measured signals themselves should be used to learn atoms' waveforms, through dictionary learning [MAI 09]. For this purpose, iterative methods have been developped that alternatively update waveforms and their associated coefficients in the signals. This type of analysis, which is well adapted to the signals of interest, should be further developed in the future [HIT 13].

6.6. Spatial representations

In order to characterize the electrical activity observed by EEG, it is interesting to associate it with the brain regions where it originates. The position of the different electrodes can be used for this: primary visual phenomena may be found on occipital electrodes and motor phenomena on central electrodes. The spatial localization of electrical potential on the scalp provides coarse information about the origin of electrical activity, called a *topographical representation* (section 6.6.1). For a more detailed spatial analysis, it is common to instead rely on a spatial filtering or localisation (sections 6.6.2 and 6.6.3).

6.6.1. *Topographic representations*

With a monopolar montage (potential differences between each sensor and a common reference), and a so-called "high (spatial) resolution" EEG, the distribution of electrodes on the scalp provides a map of the electrical potential, much like a temperature map of the globe. In fact, in order to obtain such a representation, it is necessary to interpolate the data between the measurement electrodes. This interpolation can be done with spherical splines (which assume the electrodes to be distributed on a sphere), or with two-dimensional splines on a mesh matching the shape of the scalp, which is a better approximation [PER 87].

How many electrodes are required for a "high-resolution EEG"? With 20 electrodes, it is possible to obtain a coarse topographic representation, and 64 electrodes or more produce a fairly accurate continuous map of potential. It is not reasonable to exceed 256 electrodes (the densest system currently on the market) for two reasons. On the one hand, it would be difficult to obtain measurements without creating isopotentials (in cases where electrodes are placed using conductive gel), and on the other hand that number is sufficient to properly represent the potential field present on the scalp. Indeed, due to the low conductivity of the skull, which stands in between sources and sensors, the electrical potential has rather low spatial frequencies on the scalp [NUN 06]. According to Shannon's theory, it can be accurately represented with a low-density spatial sampling. In contrast to EEG, magnetoencephalography has higher spatial frequencies and several hundred sensors are useful to accurately represent the magnetic field.

6.6.2. *Spatial filtering*

The electrical potential measured by EEG results from the superposition of different sources of cerebral activity, which create variations in electric potential on the scalp through volume conduction. Separating these different sources is difficult, but worthwhile because it yields signals that are closer to the underlying electrical brain activity. Spatial filtering offers this possibility of "deconvolving" the sensor measurements, like a pair of glasses compensates for short sightedness.

Before presenting the possible strategies, we should mention that the source mixture is instantaneous (Maxwell quasistatic). Consider M brain sources of amplitude $\jmath_m$. For each time sample n, the potential measured at each sensor k is of the form[3] $f_k[n] = \sum_{m=1}^{M} g_{km}\jmath_m[n]$. This relationship is a linear one that can be written as a matrix–vector product $f = \mathbf{G}\,J$, where the matrix $\mathbf{G}$ is called *gain matrix*. Each element g_{km} of the gain matrix represents the potential on the sensor indexed by k resulting from the presence of a unitary source (of magnitude 1) at the position of the source indexed by m.

3 In this model, measurement noise is ignored.

6.6.2.1. *Surface Laplacian*

The electric potential, denoted V, satisfies the relation:

$$\nabla \cdot (\sigma \nabla V) = \nabla \cdot \mathbf{J}^{\mathrm{p}} \qquad [6.4]$$

where σ denotes the conductivity of the head and $\mathbf{J}^{\mathrm{p}}$ are the primary sources. In areas where there are no primary electrical sources (e.g. skull and scalp), $\nabla \cdot (\sigma \nabla V) = 0$. So the Laplacian of the potential vanishes in those regions: $\Delta V = 0$.

On a surface S with normal vector $\mathbf{n}$, the surface Laplacian $\Delta_S V$ is connected to the volumic Laplacian by $\Delta V = \Delta_S V + \frac{\partial^2 V}{\partial \mathbf{n}^2}$. On the scalp, the surface Laplacian provides an interesting spatial filter because $\Delta_S V$ is close to the normal current on the skull. Indeed, as the Laplacian cancels out within the scalp, we have $\Delta_S V = -\frac{\partial^2 V}{\partial \mathbf{n}^2}$. But since air is non-conductive, by normal current conservation, on the scalp we have $\frac{\partial V}{\partial \mathbf{n}} = 0$, so $-\frac{\partial^2 V}{\partial \mathbf{n}^2}$ is approximately proportional to the normal derivative of potential on the skull. To compute the surface Laplacian, we must know V on the scalp, as well as the shape of the scalp's surface. The "surface Laplacian" most commonly used in BCI, which simply subtracts from a channel the average of its neighbors, is a very coarse approximation of the actual $\Delta_S V$, which simply subtracts the average of its neighbors from each channel. Figure 6.2 shows several possible choices for the neighbors, the "large" (respectively "narrow") Laplacian filters being sensitive to deep (respectively superficial) cortical sources.

6.6.2.2. *Cortical current density*

We have seen that the surface Laplacian is close to the normal current on the skull. Some techniques proceed even further to reconstruct the normal current on the cortical surface, which is even closer to the primary sources. This is possible because, in areas where there are no sources (the scalp, the skull), the potential is harmonic ($\Delta V = 0$), and mathematically, there is existence and uniqueness of the continuation of a function which is harmonic within a domain. However, the problem is "ill-posed" because the solution of this continuation problem is unstable: small changes to data on the boundary can lead to major changes in the function within the domain. But regularized numerical approximations have been proposed, which use, for example, finite boundary elements [HE 02, CLE 07].

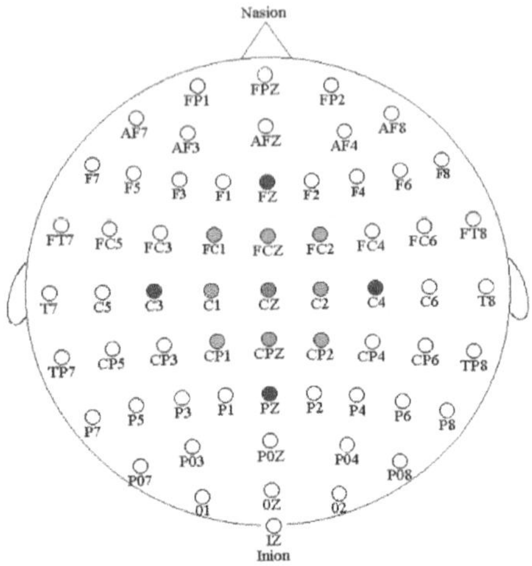

Figure 6.2. *Three Laplacian filters applied to Cz. The average of the signal measured at C3, C4, Fz and Pz (blue) is withdrawn to the Cz channel (red) for wide Laplacian filtering. The electrodes used for a medium and narrow Laplacian filter are green and orange, respectively. For a color version of this figure, see www.iste.co.uk/clerc/interfaces1.zip*

6.6.3. *Source reconstruction*

Beyond the normal current density, it can be even more interesting to identify the distribution of primary sources $\mathbf{J}^{\mathrm{p}}$ responsible for the observed measurements. The source reconstruction problem is ill-posed: it is unstable with respect to measurements, and in some cases the distribution of sources may be non-uniquely determined by the boundary measurements. However, mathematical results from potential theory guarantee uniqueness in the presence of exact measurements on a dense (continuous) portion of the boundary. The sources should obey one of the two models below (Figure 6.3):

– sources distributed orthogonally to the surface of the cortex: the brain activity that can be measured on EEG comes from pyramidal neurons (Chapter 3) and it can be represented by an extended source on a surface S, oriented in the normal direction $\mathbf{n}(\mathbf{r})$ at each position $\mathbf{r}$ of S, and whose amplitude is denoted $j(\mathbf{r})$:

$$\mathbf{J}^{\mathrm{p}}(\mathbf{r}) = j(\mathbf{r})\mathbf{n}(\mathbf{r})\delta_S(\mathbf{r})$$

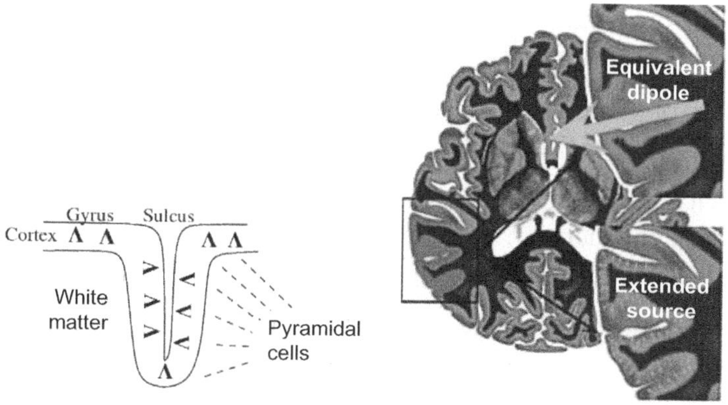

Figure 6.3. *Left: Cortical activity is locally perpendicular to the cortical surface. Right: according to the resolution at which it is described, the same cortical activity can be represented by an extended source (top inset) or an equivalent dipole. For a color version of this figure, see www.iste.co.uk/clerc/interfaces1.zip*

Since $\mathbf{n}(\mathbf{r})$ is known, source reconstruction consists of estimating the distribution of amplitudes $j(\mathbf{r})$ on the cortical surface:

– sources formed by a superposition of single dipoles:

$$\mathbf{J}^{\mathrm{p}}(\mathbf{r}) = \sum_{m=1}^{M} \mathbf{q}_m \delta_{\mathbf{p}_m}(\mathbf{r})$$

This model is acceptable when the activity is quite focal, possibly with several (M) foci. Source reconstruction consists of estimating the number M of dipoles, their position $\mathbf{p}_m$ and moment $\mathbf{q}_m$ (a vector representing amplitude and orientation).

Chapter 2 has already mentioned and illustrated source localization. In a nutshell, the sources J are estimated by minimizing a quantity of the form $||f - \mathbf{G}\,J||$ measuring a distance between measurements and their prediction, assuming the source were J. Source reconstruction requires precise knowledge of the gain matrix $\mathbf{G}$, mentioned in section 6.6.2, which connects the brain sources to sensor measurements. To calculate the gain matrix, it is necessary to solve the forward problem [6.4], and therefore, model the electrical conductivity σ of all tissues composing the head. Three types of geometric models exist for conductivity, as shown in Figure 6.4, each corresponding to a family of numerical methods:

– models of nested spheres use numerical methods based on spherical harmonic expansions [DE 93];

– surfacic models, whose surfaces are assumed to separate tissues with homogeneous conductivity, use numerical methods based on finite boundary elements [KYB 05];

– volumic models, whose conductivity can be inhomogeneous and even anisotropic, use numerical methods based on 3D finite elements, with tetrahedral, hexahedral meshes [WAB 07] or meshless, implicit formulations [VAL 10].

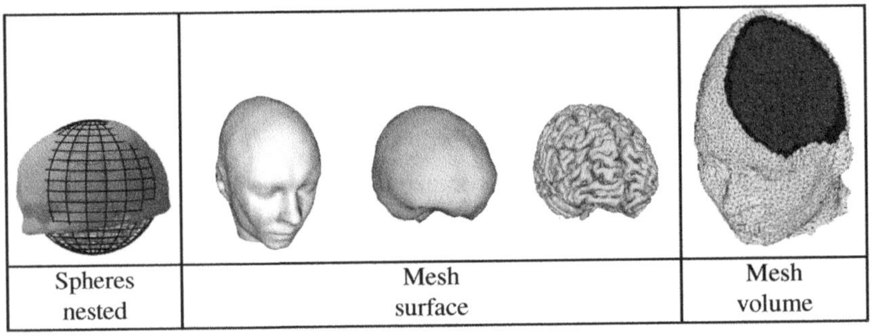

Figure 6.4. *Geometric representations of head conductivity used to compute the EEG forward problem. For a color version of this figure, see www.iste.co.uk/clerc/interfaces1.zip*

6.6.4. *Using spatial representations in BCI*

In BCI, the use of the Laplacian filter has become common, but more advanced spatial representations such as source reconstruction or cortical density are seldom used [NOI 08]. Indeed, they require geometrical data (the geometry of head tissues and electrode positioning), and their complex processing pipelines still require specific expertise. Moreover, some of these methods are still too expensive in terms of computing time to be applied online. Still, the benefit of the brain source reconstruction is to provide an excellent spatial and temporal separation of brain characteristics [BUR 15] and to produce generalizable features across subjects. This has been highlighted for error potential detection [DYS 15]. Restorative BCIs (which aim to restore efficient brain activity) should look for features in source space in order to better promote positive changes in brain activity. In the future, BCI should benefit from advances in source reconstruction methods, since this also addresses a need for better understanding the neurological mechanisms of BCI.

6.7. Statistical representations

It is sometimes difficult to identify relevant information from a set of measurements, or even from the various preprocessed representations described above. Statistical representations are useful for analyzing sets of signals measured simultaneously or through several trials (or realizations in statistical language). In Chapter 8, we will see that statistical analysis is, for example, essential for the preprocessing of extracellular potentials.

In the case of EEG, many families of statistical representations have been proposed (Bayesian methods, ergodic theory, etc.), and two of them, particularly used for in BCI, will be presented here: principal component analysis (PCA) and independent component analysis (ICA), which has already been discussed in Chapter 2.

6.7.1. *Principal component analysis*

A set of time courses measured by K channels $\{f_k[n], n = 1 \ldots N\}_{k=1 \ldots K}$ can be seen as a cloud of N points of IR^K. PCA estimates an orthogonal transformation (i.e. rotation) of $\mathbb{R}^K$ such that in the new arrangement, the axes are aligned in the direction of the point cloud's greatest dispersion. This is equivalent to diagonalizing the empirical covariance matrix of the data.

PCA must be conducted on data with a high signal-to-noise ratio (for example after averaging across several trials).

6.7.2. *Independent component analysis*

ICA is a blind source separation method. It assumes that the EEG comes from a mixture of K *statistically independent* sources s_j, for each channel $k = 1, \ldots, K$,

$$f_k[n] = \sum_{j=1}^{K} a_{kj} s_j[n]$$

The mixing coefficients can be organized in a matrix $\mathbf{A}$. ICA finds a "demixing" matrix $\mathbf{W}$ that approximates the inverse of the mixing matrix $\mathbf{W} \sim \mathbf{A}^{-1}$, so as to maximize the statistical independence of the reconstructed sources

$$\hat{s}_k[n] = \sum_{j=1}^{K} w_{kj} f_j[n]$$

A heuristic for approaching the mutual independence of signals is to search for $\mathbf{W}$ that maximizes the non-Gaussianity of signals $\hat{s}_k$, or that minimizes the mutual information between the $\hat{s}_k$ [HYV 00].

6.7.3. *Using statistical representations in BCI*

PCA requires data with a high signal to noise ratio, which, in practice, has already been averaged over several trials so for BCI it can be applied to training data. But the PCA rotation, once estimated, can be applied online to new data, in order to enhance it.

The same principle as PCA is used in the common spatial patterns (CSP) algorithm for binary classification (see Chapter 7). The CSP algorithm operates a joint diagonalization of the covariance matrices of each of the two classes, with the aims to maximize the distance between their covariance. Recent work has focused on information geometry in the space of symmetric positive definite matrices [CON 15]. Robust detection algorithms have been derived with this approach; for instance the "Riemannian Potato" for detecting artifacts [BAR 13].

ICA extracts components from EEG signals that can sometimes be independent of brain activity and hence considered artifacts.

Electrooculographic (EOG) activity is particularly well eliminated by ICA. For this, the demixing matrix is estimated from a prerecorded dataset. EOG components are identified by their frontal topography and time course. These components are subtracted through a projection, which consists of applying the demixing matrix, canceling out EOG components and finally reapplying the mixing matrix.

The methods described in this section are limited to two dimensions (sensors $\times$ time). More sophisticated methods can operate in more dimensions, e.g. multivariate analysis with PARAFAC in three dimensions (sensors $\times$ time $\times$ frequency) [MIW 04]. Mixed effects methods can be used to separate different signal variability sources [HUA 08], with applications in BCI detection [SPI 15].

6.8. Conclusions

In BCIs, preprocessing can be applied either offline (to analyze a prerecorded dataset) or online (directly during BCI interaction).

All forms of preprocessing described above are notably applied offline in order to check that the information required for BCIs is present in the signals and to specify the features to be exploited. Preprocessing in the temporal (or time–frequency), spatial and statistical domains each employs complementary principles, and they are therefore often combined. It is quite common to reconstruct dipolar sources from cross-trial averaged signals; to apply a time–frequency analysis to reconstructed distributed sources; to perform ICA and then to analyze the topographies of different components to localize corresponding sources in the brain.

The order in which processing is performed can sometimes have an effect, but not when the processing is linear, for example applying a Laplacian filter before or after cross-trial averaging produces the same results.

This chapter has presented an overview of useful forms of preprocessing for BCI and explained the basics of its main methods and their limitations and extensions. Of course, these methods need not be confined to preprocessing, since they may more generally be used as information extraction methods for EEG signals. It is certainly also interesting to apply them retrospectively to interpret brain activity modifications occurring before, during and after BCI use.

6.9. Bibliography

[BAR 13] BARACHANT A., ANDREEV A., CONGEDO M., "The Riemannian potato: an automatic and adaptive artifact detection method for online experiments using Riemannian geometry", *TOBI Workshop IV*, Sion, Switzerland, pp. 19–20, January 2013.

[BAB 08] BABILONI F., CINCOTTI F., CARDUCCI F. *et al.*, "Spatial enhancement of EEG data by surface laplacian estimation: the use of magnetic resonance imaging-based head models", *Clinical Neurophysiology*, vol. 112, pp. 724–727, 2008.

[BRO 91] BROCKWELL P.J., DAVIS R.A., *Time Series: Theory and Methods*, Springer Series in Statistics, Springer-Verlag, New York, 1991.

[BÉN 09] BÉNAR C., PAPADOPOULO T., TORRÉSANI B. *et al.*, "Consensus matching pursuit for multi-trial EEG signals", *Journal of Neuroscience Methods*, vol. 180, pp. 161–170, 2009.

[BRI 01] BRILLINGER D.R., *Time Series: Data Analysis and Theory*, Classics in Applied Mathematics, Society for Industrial and Applied Mathematics, Philadelphia, 2001.

[BUR 15] BURLE B., SPIESER L., ROGER C. *et al.*, "Spatial and temporal resolutions of eeg: is it really black and white? a scalp current density view", *International Journal of Psychophysiology*, vol. 97, no. 3, pp. 210–220, 2015.

[CON 15] CONGEDO M., AFSARI B., BARACHANT A., "Approximate joint diagonalization and geometric mean of symmetric positive definite matrices", *PLoS ONE*, vol. 10, no. 4, p. e0121423, 2015.

[CLE 07] CLERC M., KYBIC J., "Cortical mapping by laplace-cauchy transmission using a boundary element method", *Inverse Problems*, vol. 23, pp. 2589–2601, 2007.

[CLE 16] CLERC M., BOUGRAIN L., LOTTE F., *Brain–Computer Interfaces 2: Technology and Application*, ISTE, London and John Wiley and Sons, New York, 2016.

[DE 93] DE MUNCK J.C., PETERS M.J., "A fast method to compute the potential in the multisphere model", *IEEE Transactions on Biomedical Engineering*, vol. 40, no. 11, pp. 1163–1174, 1993.

[DYS 15] DYSON M., THOMAS E., CASINI L. *et al.*, "Online extraction and single trial analysis of regions contributing to erroneous feedback detection", *NeuroImage*, vol. 121, pp. 146–158, 2015.

[FLA 93] FLANDRIN P., *Temps-Fréquence*, Hermès, Paris, 1993.

[GRI 08] GRIBONVAL R., RAUHUT H., SCHNASS K. *et al.*, "Atoms of all channels, unite! average case analysis of multi-channel sparse recovery using greedy algorithms", *The Journal of Fourier Analysis and Applications*, vol. 14, no. 5, pp. 655–687, 2008.

[HE 02] HE B., ZHANG X., LIANG J. *et al.*, "Boundary element method-based cortical potential imaging of somatosensory evoked potentials using subjects' magnetic resonance images", *NeuroImage*, vol. 16, pp. 564–576, 2002.

[HIT 13] HITZIGER S., CLERC M., GRAMFORT A. *et al.*, "Jitter-adaptive dictionary learning – application to multi-trial neuroelectric signals", *ICLR – 1st International Conference on Learning Representations*, May 2013.

[HUA 08] HUANG Y., ERDOGMUS D., PAVEL M. *et al.*, "Mixed effects models for eeg evoked response detection", *IEEE Workshop on Machine Learning for Signal Processing, MLSP*, pp. 91–96, 2008.

[HYV 00] HYVÄRINEN A., OJA E., "Independent component analysis: algorithms and applications", *Neural Networks*, vol. 13, pp. 411–430, 2000.

[KYB 05] KYBIC J., CLERC M., ABBOUD T. *et al.*, "A common formalism for the integral formulations of the forward eeg problem", *IEEE Transactions on Medical Imaging*, vol. 24, pp. 12–28, 2005.

[MAL 08] MALLAT S., *A Wavelet Tour of Signal Processing*, Academic Press, New York, 2008.

[MIW 04] MIWAKEICHI F., MARTINEZ-MONTES E., VALDÉS-SOSA P.A., *et al.*, "Decomposing EEG data into space-time-frequency components using parallel factor analysis", *NeuroImage*, vol. 22, no. 3, pp. 1035–1045, 2004.

[MAI 09] MAIRAL J., PONCE J., SAPIRO G. *et al.*, "Supervised dictionary learning", *Advances in Neural Information Processing Systems*, pp. 1033–1040, 2009.

[MAL 93] MALLAT S., ZHANG Z., "Matching pursuit with time-frequency dictionaries", *IEEE Transactions on Signal Processing*, vol. 41, no. 12, pp. 3397–3414, 1993.

[NOI 08] NOIRHOMME Q., KITNEY R., MACQ B., "Single-trial EEG source reconstruction for brain–computer interface", *IEEE Transactions on Biomedical Engineering*, vol. 55, no. 5, pp. 1592–1603, 2008.

[NUN 06] NUNEZ P., SRINIVASAN R., *Electric Fields of the Brain*, Oxford University Press, Oxford, New York, 2006.

[PER 87] PERRIN F., BERTRAND O., PERNIER J., "Scalp current density mapping: value and estimation from potential data", *IEEE Transactions on Biomedical Engineering*, vol. 34, no. 4, pp. 283–288, 1987.

[SPI 15] SPINNATO J., ROUBAUDM-C., BURLE B. *et al.*, "Detecting single-trial EEG evoked potential using a wavelet domain linear mixed model: application to error potentials classification", *Journal of Neural Engineering*, vol. 12, no. 3, p. 036013, 2015.

[SPU 15] SPUSTEK T., JEDRZEJCZAK W., BLINOWSKA K., "Matching Pursuit with Asymmetric Functions for Signal Decomposition and Parameterization", *PloS one*, vol. 10, no. 6, 2015.

[VAL 10] VALLAGHÉ S., PAPADOPOULO T., "A trilinear immersed finite element method for solving the electroencephalography forward problem", *SIAM Journal on Scientific Computing*, vol. 32, no. 4, pp. 2379–2394, 2010.

[WAB 07] WOLTERS C.H., ANWANDER A., BERTI G. *et al.*, "Geometry-adapted hexahedral meshes improve accuracy of finite-element-method-based eeg source analysis", *IEEE Transactions on Biomedical Engineering*, vol. 54, no. 8, pp. 1446–1453, 2007.

7

EEG Feature Extraction

7.1. Introduction

One of the major steps in the design of a BCI that uses EEG signals is the processing and classification of these signals to identify the user's mental state. As previously seen, EEG signal analysis is divided into three stages: preprocessing (discussed in Chapter 6), feature extraction and classification. This chapter focuses on feature extraction, which consists of describing EEG signals by (an ideally small) core set of values describing the relevant information they contain in order to later classify them. In particular, we will see what type of information to extract from EEG signals for different types of BCIs, and how to extract this information in order for it to best discriminate between different mental states.

7.2. Feature extraction

A feature is a value that describes a property of EEG signals, for example the power of the EEG signal in the μ rhythm for the C3 electrode. The characteristics are usually grouped together in a vector called a "feature vector". As an example, let us look at a BCI using motor imagery (MI) – that is to say that it can recognize imagined movements, for example of the hands. In this case, the two mental states to be identified are imagined movements of the left hand, and imagined movements of the right hand. The feature that is

---

Chapter written by Fabien LOTTE and Marco CONGEDO.

usually extracted in order to identify these mental states in the EEG signals is band power, i.e. the EEG signal's strength in a specific frequency band. For MI, band powers are usually calculated in μ (about 8–12 Hz) and β (about 16–24 Hz) frequency bands for electrodes located over the sensorimotor cortex (e.g. C3 and C4 electrodes for imagined hand movements) [PFU 01]. Such features can typically be classified using linear discriminant analysis (LDA, see Chapter 8).

In BCI design, EEG signal processing often relies on machine learning techniques. This means that the classifier and/or features are usually adjusted and optimized for each user, using examples of EEG signals produced by each user. These examples of EEG signals are called a training set and are labeled with their membership class – that is to say the user's mental state when the EEG signals were recorded. With this training set, it becomes possible to calibrate a classifier capable of recognizing the class of different EEG signals, as will be described in Chapter 8. Features can also be optimized through examples of EEG signals, for example by selecting the most relevant electrodes to recognize different mental states. Thus, designing a BCI that employs machine learning (which is the case of most BCIs) requires (1) a calibration phase (also called a training phase) that consists of acquiring training EEG signals (i.e. examples) and optimizing the EEG signal processing chain by adjusting the feature settings and/or by adjusting a classifier and (2) a usage phase (also called a test phase), which consists of using the model (features and classifier) obtained during calibration in order to recognize the user's mental state based on new EEG signals (i.e. EEG signals other than those in the training set) to operate the BCI.

As briefly mentioned here, and as we shall see in detail in Chapter 8, a classifier is able to learn what class corresponds to which input features by examining examples. So why not use the values of EEG signals directly as an input to a classifier? The answer is that would most likely not work because of a phenomenon called the "curse of dimensionality": it has been observed in practice that the amount of examples needed to properly describe different classes increases exponentially with the dimension of the feature vector (i.e. the number of features used) [RAU 91]. Some researchers even recommend using 5–10 times more training examples than the size of the feature vector[1] [RAU 91]. What would this recommendation mean if we used the value of EEG signals as features directly? For example, suppose we were using a common EEG system with 32 electrodes sampled at 250 Hz, and an EEG signal example for a given mental task lasts 1 s. We would therefore have a feature vector with a size of $32 \times 250 \times 1 = 8,000$, which would require at least $8,000 \times 5 = 40,000$ training examples by class. Obviously, we cannot ask a BCI user to perform each mental task $40,000$ times in order to calibrate the BCI before he/she can use it. So we need a more compact representation, and thus extract features of EEG signals.

In BCI design, there are three main sources of information that can be used to extract EEG signal features:

– *spatial information*: this describes where (spatially) the relevant signal comes from. In practice, this means selecting specific EEG electrodes or focusing more on some than on other sensors. In other words, this is equivalent to using the signal from one or several specific brain regions, but not (or very little) from any others;

– *spectral (or frequency) information*: this describes how the power of the EEG signal varies in some specific frequency bands. In practice, this is equivalent to using signal band power as features.

– *temporal information*: this describes how EEG signals vary over time. In practice, this means using the values of EEG signals for different specific time intervals (though not all) or different time windows.

It is usually necessary to use different sources of information for different types of BCIs. In particular, BCIs using oscillatory EEG activity (e.g. BCIs based on MI) primarily use spatial and spectral information, while BCIs employing evoked potentials (EP) mainly use spatial and temporal information. The following sections describe feature extraction techniques used for these two types of BCIs.

1 This recommendation was made before the invention of SVM (described in Chapter 8), which are less sensitive to the curse of dimensionality.

7.3. Feature extraction for BCIs employing oscillatory activity

BCIs employing oscillatory activity use mental states that produce amplitude changes in EEG oscillations, that is to say changes in the power of EEG signals in certain frequency bands (ERD/ERS, see Chapter 4). Such BCIs include most notably BCIs employing MI [PFU 01], SSVEP [VIA 10], various mental imagery tasks such as mental arithmetic or mental generation of words [FRI 12] and even different levels of mental workload [MÜH 14].

This section first presents a basic (and overly simplistic) description of that kind of BCI, and then focuses on some more advanced tools including the major algorithm called "common spatial patterns" (CSP).

7.3.1. *Basic design for BCI using oscillatory activity*

BCIs that employ oscillatory activity use the power of the EEG signals in certain frequency bands (spectral information) and certain brain regions (spatial information).

For example, the basic design of an MI-based BCI would use spatial information by extracting features only from EEG sensors located above the brain motor areas, typically C3 sensors for imagined movement of the right hand, Cz for imagined movement of the feet and C4 for imagined movement of the left hand. It would use the spectral information focusing on frequency bands μ (8–12 Hz) and β (16–24 Hz). Specifically, for a BCI that can recognize imagined movements of the left and right hands, basic features would be band power in 8–12 Hz and 16–24 Hz for each of electrodes C3 and C4, which thus constitutes four features. There are many ways to calculate EEG signal band power [BRO 11]. However, a simple but effective technique is to first filter the EEG signal of a given sensor in the target frequency band, and then to calculate the square of the filtered signal to obtain its power, finally averaging the signal over time (e.g. during the last second, using sliding windows). This principle is shown in Figure 7.1.

Unfortunately, this basic design is far from optimal. Indeed, it uses only two electrodes: relevant information measured by other sensors may be missing. In addition, C3 and C4 are perhaps not the best sensors for the subject being studied. Similarly, the fixed frequency bands at 8–12 Hz and 16–24 Hz are perhaps not the best bands for the subject at hand. In general, better performance can be obtained using a specific design for each subject by optimizing the best electrodes and frequency bands. Using more than two sensors also makes it possible to obtain better performance.

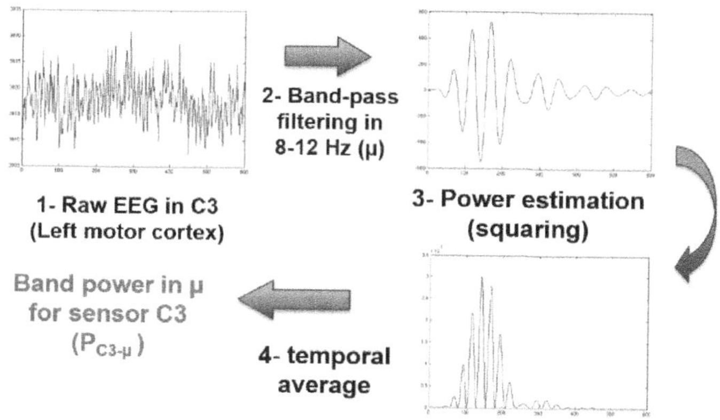

Figure 7.1. *Band power feature extraction from a raw EEG signal. The EEG signal illustrated here was recorded during an instance of imagined right hand movement (beginning at $t = 0$ s). The contralateral ERD during the imagined movement is clearly visible: the signal power of the C3 electrode (left motor cortex) in 8–12 Hz band clearly decreases during imagined movement*

7.3.2. *Toward more advanced, multiple electrode BCIs*

The need to use more than two sensors as well as specific sensors in each subject has lead to the design of BCIs that use multiple electrodes. This need has been confirmed by various studies that suggest that, for MI, maximum performance is achieved with a large number of electrodes, for example with 48 sensors according to [SAN 10]. However, simply adding sensors will not solve performance issues. In fact, using more sensors means extracting more features, which makes the curse of dimensionality even more likely. Therefore, just adding sensors can even reduce performance if the number of available training examples is too low. In order to efficiently employ multiple sensors, there are three main approaches, each contributing to reducing the dimension of the feature vector:

– *feature selection algorithms*: these are methods for automatically selecting a relevant subset of features among the features initially extracted [GUY 03];

– *sensor selection algorithms*: these are methods similar to those employed in feature selection and are aimed at automatically selecting a relevant subset of sensors from among all available sensors;

– *spatial filtering algorithms*: these are methods for combining several sensors, generally through linear combination, in order to form a new (virtual) sensor from the extracted features.

Later, we will focus on spatial filtering, for which algorithms specific to EEG and BCIs have been developed. Feature selection is indeed a set of general tools in machine learning, which is not specific to EEG or BCI (see [GUY 03] for details). As for sensor selection, the algorithms used are usually derived from the feature selection algorithms. Readers interested in the topic can refer, for example, to [SCH 05] or [ARV 11] and their references to learn more about the subject.

7.3.2.1. *Spatial filtering*

Spatial filtering consists of using a small number of new channels (virtual channels) that are defined as a linear combination of the original sensors. Formally, spatial filtering is described by $\tilde{x} = \sum_i w_i x_i = wX$, where $\tilde{x}$ is the spatially filtered signal, x_i is the EEG signal from sensor i, w_i is the weight given to this sensor in the spatial filtering and X is a matrix whose ith row is x_i, i.e. X is the EEG signal matrix for each sensor. Spatial filtering is useful not only because it reduces the size of the problem (passing from many initial EEG sensors to a small number of spatially filtered signals – far fewer spatial filters than the number of original sensors are typically used) but also because it has a neurophysiological meaning. Indeed, as discussed above (Chapters 2 and 3), EEG signals measured on the surface of the scalp are the result of noisy mixtures of EEG signals from different brain regions. In other words, since the EEG signal from the cortex is diffused as it passes through the skull or scalp, when it arrives at the EEG sensors, said signal is diffused and dispersed on several EEG sensors. Therefore, spatial filtering makes it possible to help recover the original signal (from the cortex) by gathering relevant information that was scattered over different sensors.

There are different ways to define the spatial filters. In particular, the weight of a filter w_i can be fixed beforehand, according to neurophysiological knowledge, or it can be optimized using training examples. Among the fixed spatial filters, we can mention bipolar and Laplacian filters in particular, which have already been described in the previous chapter. It has been shown that extracting features from bipolar or laplacian channels rather than from the original EEG sensors significantly increases classification performance [MCF 97]. Methods for reconstruction of distributed sources (also presented in the previous chapter) can also be used to define fixed spatial filters in order to analyze the EEG signal from very specific brain regions. Extracting features from spatial filter obtained by source reconstruction also allows for better classification performance than when extracting features from the original EEG sensors [CON 06].

The second category of spatial filters, which are based on data, contains filters that are optimized for each subject on the training data. This category contains in particular the spatial filters constructed through independent component analysis (ICA) [KAC 08], as described in the previous chapter. These techniques make it possible to obtain the weight w_i of spatial filters in an unsupervised manner, that is to say without knowing the labels (classes) of the training data. Alternatively, the weight of the spatial filters can be defined in a supervised manner (i.e. knowing the label of each training example) in order to optimize a measure of separation between classes. One such algorithm has been developed for BCI based on EEG oscillatory activity: the CSP algorithm [RAM 00], which is described below.

7.3.3. *The CSP algorithm*

Informally, the CSP algorithm optimizes spatial filters w such that the variance of the filtered EEG signal is maximum for one class and minimal for another class. Since the variance of a filtered signal in the frequency band b is equal to the power of the signal in the band b, this means that CSP optimizes the spatial filters to obtain the band power features that are optimally discriminant because their value is maximally different between the two classes. CSP is therefore especially useful for BCIs based on oscillatory activity, since the most useful type of features for their design is precisely band power. For example, for BCIs employing MI, EEG signals are typically filtered in the 8–30 Hz band (μ and β rhythms) before being spatially filtered by CSP [RAM 00]. Formally, CSP optimizes the spatial filters w by extremizing (i.e. minimizing and maximizing) the following function:

$$J_{CSP1}(w) = \frac{wX_1X_1^Tw^T}{wX_2X_2^Tw^T} \qquad [7.1]$$

which is equivalent to extremizing

$$J_{CSP2}(w) = \frac{wC_1w^T}{wC_2w^T} \qquad [7.2]$$

where T is the matrix transpose, X_i are EEG training signals for class i, which were previously bandpass filtered (matrix with EEG samples as columns and sensors as rows) and C_i the spatial covariance matrix for class i. In practice, the covariance matrix C_i is defined as the average of the covariance matrices of each example of the class i [RAM 00]. In this equation, wX_i is the spatially filtered EEG signal for class i, and $wX_iX_i^Tw^T$ is therefore the variance of the spatially filtered signal, that is to say its band power. So maximizing and minimizing $J_{CSP}(w)$ makes it possible to obtain spatially filtered signals whose band power is maximally different between classes. $J_{CSP}(w)$ is in the form of what is mathematically called a generalized Rayleigh quotient. Therefore, maximizing and minimizing this function can be solved by Generalized Eigen Value Decomposition (GEVD). The spatial filters w that maximize or minimize $J_{CSP}(w)$ are therefore the eigenvectors corresponding to the largest and smallest values of the GEVD of matrices C_1 and C_2. Typically, six CSP filters are used (i.e. three pairs of filters), which correspond to the three largest and three smallest eigenvalues. Once the filters have been obtained, a CSP feature f is calculated as follows:

$$f = \log(wCw^T) \approx \log(var(wX)) \qquad [7.3]$$

i.e. a CSP feature is simply the band power of the signal spatially filtered with one of the CSP filters w. The use of CSP is shown in Figure 7.2. In this figure, a marked difference in variance (and hence band power) can be observed between signals spatially filtered with CSP for each of the two classes, which provides good classification performance.

The CSP algorithm has many advantages: first, it provides relatively high classification performance for BCIs. This is a fairly flexible algorithm, since it can be used for any BCI using the ERD/ERS. Finally, it is numerically efficient and a simple algorithm to implement. All of this means that CSP has become one of the most popular and efficient algorithms for designing BCIs based on oscillatory activity [LOT 11]. Their performance is illustrated in the following section.

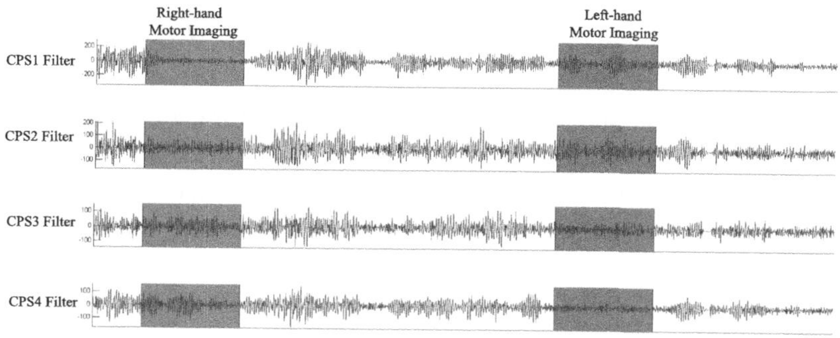

Figure 7.2. *EEG signals spatially filtered with the common spatial patterns (CSPs) algorithm. The first two spatial filters (both signals from the top) are those that maximize the variance of the signals in the "imagined movement of the left hand" class (in red) while minimizing those of the "imagined movement of the right hand" class (in green). The last two filters (the two signals on the bottom) do the opposite, i.e. they maximize the variance of the "imagined movement of the right hand" class, while minimizing the variance of the "imagined movement of the left hand" class. For a color version of this figure, see www.iste.co.uk/clerc/interfaces1.zip*

7.3.4. *Illustration on real data*

In order to illustrate the impact of spatial filters on BCI classification performance, we compared the performances obtained by different filters on the dataset IIA from "BCI competition IV" [TAN 12]. During this competition, a training set and a test set were made available to competitors. Each set contains 72 examples of EEG signals for each class (imagining a movement of the left hand, right hand, tongue and feet) for nine different subjects. The training set examples were labeled with their class, while the examples for the test set were not labeled: competitors had to calibrate their algorithms (e.g. the classifier) on the training set, and use it to guess the class of test examples. The aim of the competition was to identify the best algorithms for recognizing the different mental tasks. Data from this competition is now available for testing, evaluating and comparing different methods.

We compared four different BCI designs offline, each based on one of four different spatial filters and aimed at distinguishing imagined movements of the right and left hands. The four filters used were (1) no filter, just EEG signals from electrodes C3 and C4; (2) bipolar filter around C3 and C4; (3) Laplacian filter around C3 and C4 and (4) CSP filters (three pairs of filters). In order to classify the signals, we first performed a spectral filtering in the 8–30 Hz band, and then filtered spatially with one of the four filters. We then calculated the signal's average band power over a time window of 2 s starting 0.5 s after the beginning of the imagined task. These band powers for each filter are then given to a classifier (LDA) to identify the task performed. CSP and LDA are calibrated on the training set, and each method is then tested on the test set. Figure 7.3 shows the classification performance obtained on the test set by each method.

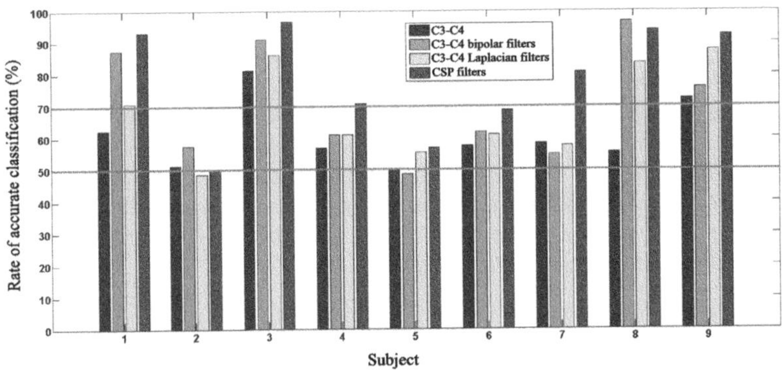

Figure 7.3. *Classification performance (rate of accurate classification) obtained on data set IIa for "BCI competition IV" [TAN 12] in classifying imagined movement of the right hand using different spatial filters. For a color version of this figure, see www.iste.co.uk/clerc/interfaces1.zip*

As can be seen in this figure, using a fixed spatial filter, such as bipolar or Laplacian filters, makes it possible to increase performance (average performance for bipolar filters: 70.5%, for Laplacian filters 68%) as compared to using only C3 and C4 (average performance 60.7%). Using a data-based filter such as CSP will increase classification performance even more (average performance of 78.1%). It should, nevertheless, be noted that CSP is not a perfect algorithm, far from it. It is especially sensitive to the presence of noise and artifacts, and does not work well with little training data. Many variations of CSP have been proposed to remedy this, and research on the subject is still very active [LOT 11, SAM 14].

7.4. Feature extraction for the BCIs employing EPs

A typical example of an EP used in BCIs is P300 [FAZ 12], as described in Chapter 4. EPs are characterized by specific temporal variations appearing in response to a stimulus. Thus, unlike BCIs using brain activity, BCIs using the EP mainly employ temporal information rather than spectral information. Nevertheless, such as BCI employing oscillatory activity, those exploiting EP can also exploit spatial information.

For example, in the case of a BCI using P300, spatial information is used that focuses mainly on parietal and occipital electrodes (i.e. by extracting features from these electrodes only), where P300 comes from. Krusienski *et al.* recommend, for example, using a set of eight sensors, located in Fz, Cz, P3, Pz, P4, PO7, Oz and PO8 [KRU 06]. Once the relevant spatial information has been identified – i.e. the electrodes (as in the example above) – features can be extracted from each of their signals. For EPs in general and therefore also for P300, the features used usually reflect the signals' temporal information, that is to say how the signal amplitude varies with time. This is done by using the value of the different points of EEG signals preprocessed as features. More specifically, features are usually extracted from an EP (1) by performing a low-pass or bandpass filtering of the signals (e.g.in 1–12 Hz for P300, given that EPs are usually slow waves), (2) by subsampling the filtered signal to reduce the number of EEG signal points and therefore the dimension of the problem, and (3) by collecting the values of the remaining EEG points for all selected sensors into a single feature vector which will be used as the input of a classifier. This process is illustrated in Figure 7.4 in the extraction of features from the Pz electrode for a BCI using P300.

Extracting several time points from several sensors as features makes BCIs employing EP usually have a higher dimension than those using oscillatory activity. Therefore, it is important to use classifiers that can handle high dimension (see the following section), or to use, as mentioned above, feature or sensor selection algorithms, which are the same for the two types of BCIs. Spatial filters devoted to EP have also been proposed.

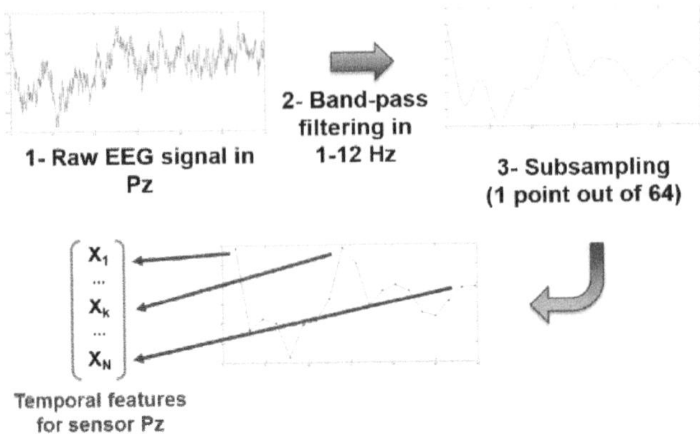

Figure 7.4. *Traditional process for extracting features from an EEG sensor for a BCI using EP (here P300). The stimulus that can evoke the EP appears at t = 0 s*

7.4.1. *Spatial filtering for BCIs employing EPs*

As with BCIs employing oscillatory activity, BCIs using the EP can also benefit from spatial filtering to identify brain sources whose features are discriminatory. So why not also use the CSP for EPs? This is due to the fact that one crucial bit of information for classifying EPs is information on EEG signals' time course. Unfortunately, CSP completely ignores this information because it only considers average signal power (and therefore not its time course) in optimizing filters. Thus, CSP is not suitable for EP classification. Fortunately, there are other spatial filtering algorithms specifically for EPs. We can most prominently mention Fisher spatial filtering, which is devoted to EPs, and was proposed by Hoffman *et al.* [HOF 06], or xDAWN, proposed by Rivet *et al.* [RIV 09]. The objective of these two spatial filters is to obtain spatially filtered signals such that the EPs are more visible and more discriminable than the original EEG signals. These two methods use different objective functions to achieve this goal. We will describe xDAWN filtering to illustrate spatial filtering of EPs for BCI. The xDAWN spatial filtering algorithm, which has proved very effective for EP classification, seeks to maximize the signal to noise ratio. Informally, this means that xDAWN seeks to highlight the EP in order to make it more visible amidst the noise.

Formally, xDAWN optimizes spatial filters by maximizing the following function:

$$J_{xDAWN} = \frac{wSS^T w^T}{wXX^T w^T} \qquad [7.4]$$

where S is the estimated average time course of the EP (averaged over the number of repetitions). The average EP can be estimated more accurately using least squares, where the EPs temporally overlap (which is the case with "P300-Speller") [RIV 09]. In this equation, the numerator represents the signal, that is to say the relevant information that it seeks to highlight. Indeed, $wSS^T w^T$ is the power of the time course of the EP after spatial filtering. In contrast, in the denominator $wXX^T w^T$ is the power of all EEG signals after spatial filtering. The denominator thus contains both the signal (the target EP) and noise. Therefore, maximizing J_{xDAWN} requires simultaneously maximizing the signal – that is to say highlighting the EP – and minimizing the signal plus the noise – that is to say making noise as small as possible [RIV 09]. This approach makes it possible to significantly improve EP classification performance, especially when few training examples are available.

7.5. Alternative methods and the Riemannian geometry approach

This chapter has presented the main tools for recognizing a user's mental state based on the EEG oscillations (spectral information) and EPs (temporal information). These tools are relatively simple, widely used and effective. These are not, however, the only tools available. We can mention, for example, other methods for extracting information, such as Hjorth temporal parameters [OBE 01] or "time domain parameters" [VID 09], methods for measuring the complexity of signals [BRO 12] or methods measuring connectivity information – that is to say measuring how the various sensor signals are connected (e.g. synchronized or correlated) [CAR 14].

Recent work has explored a new approach to feature extraction for BCIs [BAR 12, CON 13]. The idea is to summarize the relevant information, whether spatial, frequential and/or temporal, for a particular point in a multidimensional space. This space is the Riemannian manifold of symmetric positive definite (SPD) matrices.

An SPD matrix is a square and symmetric matrix for which all eigenvalues are positive. For example, the spatial covariance matrices in EEG are SPD (Figure 7.5, on the left). The Riemannian manifold of SPD matrices is their native and curved space (see Figure 7.5, on the right). On this manifold, a suitable metric gives us a way to calculate the distance between any two points (Figure 7.5, on the right).

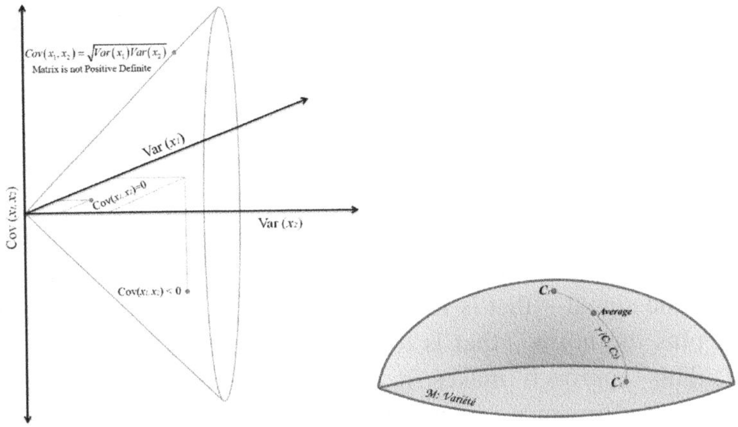

Figure 7.5. *Left: Covariance matrices with dimension $N \times N$ are constrained by their symmetry and the positive sign of their diagonal elements (variance), among others. We can easily visualize the topology in the case of 2×2 matrices; the cone on the left represents each of these matrices as a point in Euclidean space in three dimensions, where two coordinates correspond to the two variances (the two diagonal elements) and the third corresponds to the covariance (off-diagonal elements, which are equal since the matrix is symmetrical). By construction, the point must remain within the cone. As soon as that point touches the edge of the cone, the matrix is no longer SPD. Right: The geodesic going through two points C1 and C2 on the Riemannian manifold of SPD matrices is the minimum path length between them. The geometric mean of these two points is the halfway point on the geodesic, which is usually far from the arithmetic mean* $\frac{1}{2}(C1 + C2)$

With this notion of distance, we can estimate the geometric mean of a cloud of points (each point being an SPD matrix representing the EEG signal) corresponding to the different classes of a BCI obtained in the calibration phase. We can thus classify new signals simply by evaluating their distance from the geometric mean of each class (Figure 7.6). This approach proved to be as simple as it is effective, providing classification results as good as those obtained by the state-of-the art for MI and P300, while allowing a better generalization between sessions and subjects [BAR 12, CON 13]. Moreover, this strength has also helped to design BCIs that do not require calibration. For this, a database is used to initialize the BCI and the BCI then adapts to the subject during use. Moreover, with this approach, the processing chain is identical for all types of BCIs: only the definition of the point that summarizes EEG signals on the manifold changes depending on the type of BCI. For example, for BCIs based on MI spatial information is used, and for those based on the EP, temporal information is used [CON 13].

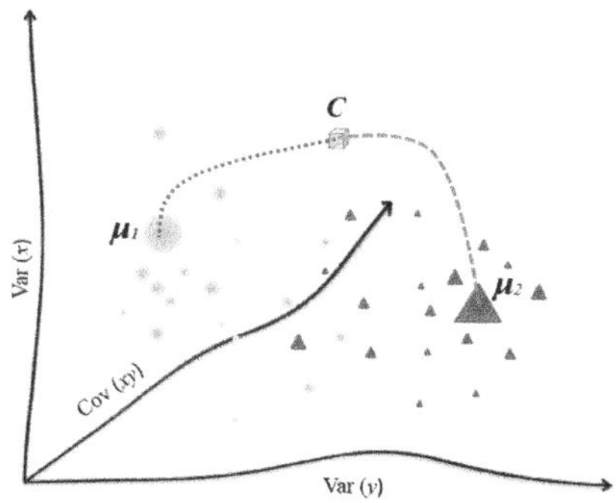

Figure 7.6. *Consider a two-class BCI with 2 × 2 matrices. With the calibration data, we estimate the geometric mean for Class 1 (μ_1) and Class 2 (μ_2). In the testing phase, a test (C) represented by an SPD matrix is classified by evaluating its distance to each geometric mean and labeled as belonging to the nearest class. The same process is used for arrays of any size and for any number of classes*

7.6. Conclusions

In this chapter, we tried to show the reader the kinds of relevant information that can be extracted from EEG signals in order to later be classified, as well as how to extract this information. In particular, we saw that the three main sources of information are (1) spectral information that is used with band power features, (2) temporal information, typically represented as the amplitude of preprocessed EEG signals in short time and (3) spatial information, which can be exploited by focusing on certain sensors or using spatial filters (CSP for BCI using oscillatory activity, or xDAWN for those operating EPs). It is important to note that there is still much work on BCI feature extraction, and the ideal method, which is both robust, invariant (over time) and universal (works well for all users), still remains to be found.

7.7. Bibliography

[ARV 11] ARVANEH M., GUAN C., ANG K. *et al.*, "Optimizing the channel selection and classification accuracy in EEG-based BCI", *IEEE Transactions on Biomedical Engineering*, vol. 58, p. 1865–1873, 2011.

[BAR 12] BARACHANT A., BONNET S., CONGEDO M. *et al.*, "Multiclass brain–computer interface classification by Riemannian geometry", *IEEE Transactions on Biomedical Engineering*, vol. 59, no. 4, pp. 920–928, 2012.

[BRO 11] BRODU N., LOTTE F., LÉCUYER A., "Comparative study of band-power extraction techniques for motor imagery classification", *Proceedings of IEEE CCMB*, pp. 1–6, 2011.

[BRO 12] BRODU N., LOTTE F., LÉCUYER A., "Exploring two novel features for EEG-based brain-computer interfaces: multifractal cumulants and predictive complexity", *Neurocomputing*, vol. 79, no. 1, pp. 87–94, 2012.

[CAR 14] CARAMIA N., LOTTE F., RAMAT S., "Optimizing spatial filter pairs for EEG classification based on phase synchronization", *Proceedings of ICASSP*, 2014.

[CON 06] CONGEDO M., LOTTE F., LÉCUYER A., "Classification of movement intention by spatially filtered electromagnetic inverse solutions", *Physics in Medicine and Biology*, vol. 51, no. 8, pp. 1971–1989, 2006.

[CON 13] CONGEDO M., EEG source analysis, PhD thesis, HDR presented at Doctoral School EDISCE, University of Grenoble, 2013.

[FAZ 12] FAZEL-REZAI R., ALLISON B., GUGER C. *et al.*, "P300 brain computer interface: current challenges and emerging trends", *Frontiers in Neuroengineering*, vol. 5, no. 14, 2012.

[FRI 12] FRIEDRICH E., SCHERER R., NEUPER C., "The effect of distinct mental strategies on classification performance for brain-computer interfaces", *International Journal of Psychophysi*ology, vol. 84, no. 1, pp. 86–94, 2012.

[GUY 03] GUYON I., ELISSEEFF A., "An introduction to variable and feature selection", *Journal of Machine Learning Research*, vol. 3, pp. 1157–1182, 2003.

[HOF 06] HOFFMANN U., VESIN J., EBRAHIMI T., "Spatial filters for the classification of event-related potentials", *Proceedings of ESANN*, 2006.

[KAC 08] KACHENOURA A., ALBERA L., SENHADJI L. *et al.*, "ICA: a potential tool for BCI systems", *IEEE Signal Processing Magazine*, vol. 25, no. 1, pp. 57–68, 2008.

[KRU 06] KRUSIENSKI D., SELLERS E., CABESTAING F. *et al.*, "A comparison of classification techniques for the P300 Speller", *Journal of Neural Engineering*, vol. 3, pp. 299–305, 2006.

[LOT 11] LOTTE F., GUAN C., "Regularizing common spatial patterns to improve BCI designs: unified theory and new algorithms", *IEEE Transactions on Biomedical Engineering*, vol. 58, no. 2, pp. 355–362, 2011.

[MCF 97] McFarland D.J., McCane L.M., David S.V. *et al.*, "Spatial filter selection for EEG-based communication", *Electroencephalographic Clinical Neurophysiology*, vol. 103, no. 3, pp. 386–394, 1997.

[MÜH 14] Mühl C., Jeunet C., Lotte F., "EEG-based workload estimation across affective contexts", *Frontiers in Neuroscience*, vol. 8, p. 114, 2014.

[OBE 01] Obermeier B., Guger C., Neuper C. *et al.*, "Hidden Markov models for online classification of single trial EEG", *Patt Rec Letters*, vol. 22, no. 12, pp. 1299–1309, 2001.

[PFU 01] Pfurtscheller G., Neuper C., "Motor imagery and direct brain–computer communication", *Proceedings of the IEEE*, vol. 89, no. 7, pp. 1123–1134, 2001.

[RAM 00] Ramoser H., Muller-Gerking J., Pfurtscheller G., "Optimal spatial filtering of single trial EEG during imagined hand movement", *IEEE Transactions on Rehabilitation Engineering*, vol. 8, no. 4, pp. 441–446, 2000.

[RAU 91] Raudys S.J., Jain A.K., "Small sample size effects in statistical pattern recognition: recommendations for practitioners", *IEEE Transactions on Pattern Analysis and Machine Intelligence*, vol. 13, no. 3, pp. 252–264, 1991.

[RIV 09] Rivet B., Souloumiac A., Attina V. *et al.*, "xDAWN algorithm to enhance evoked potentials: application to brain computer interface", *IEEE Transactions on Biomedical Engineering*, vol. 56, no. 8, pp. 2035–2043, 2009.

[SAM 14] Samek W., Kawanabe M., Muller K., "Divergence-based framework for common spatial patterns algorithms", *IEEE Review in Biomedical Eng*ineering, vol 7, pp. 50–72, 2014.

[SAN 10] Sannelli C., Dickhaus T., Halder S. *et al.*, "On optimal channel configurations for SMR-based brain–computer interfaces", *Brain Topography*, 2010.

[SCH 05] Schröder M., Lal T., Hinterberger T. *et al.*, "Robust EEG channel selection across subjects for brain–computer interfaces", *EURASIP Journal of Applied Signal Process*ing, pp. 3103–3112, 2005.

[TAN 12] Tangermann M., Müller K., Aertsen A. *et al.*, "Review of the BCI competition IV", *Frontiers in Neuros*cience, 2012.

[VIA 10] Vialatte F., Maurice M., Dauwels J. *et al.*, "Steady-state visually evoked potentials: focus on essential paradigms and future perspectives", *Progress in Neurobiology*, vol. 90, pp. 418–438, 2010.

[VID 09] Vidaurre C., Krämer N., Blankertz B. *et al.*, "Time domain parameters as a feature for EEG-based brain computer interfaces", *Neural Networks*, vol. 22, pp. 1313–1319, 2009.

8

Analysis of Extracellular Recordings

8.1. Introduction

This chapter presents the methods of analysis required for the "oldest" types of brain–machine interface. These methods are (strongly) invasive, as they require trepanation to insert a large number (10–100) of electrodes into the brain tissue. The first feasibility studies were performed on rats [CHA 99] and on monkeys [WES 00] – Chaplin [CHA 04] presents an initial overview of this work. The advantage of these methods of recording is that they give access to individual neuron activity with excellent resolution in time (achieving this resolution is the main subject of this chapter) and the obvious disadvantage is that they require trepanation, which excludes them from being used with patients, except in very exceptional cases. As the other chapters of this book will discuss, BCIs may be implemented without recourse to the invasive methods that we shall discuss here; however, these methods are still very frequently used by neurophysiologists in a wider context that we shall introduce in the following section.

Chapter written by Christophe POUZAT.

8.1.1. *Why is recording neuronal populations desirable?*

There are three main reasons why recording large numbers of neurons simultaneously while maintaining the resolution of individual neurons is desirable for neurophysiologists[1]:

1) more data collected per instance of the experiment, which limits the number of animals required for a study and reduces costs;

2) multiple models of information processing by neuronal networks, such as perceptual binding by synchronization[2], suggest that the synchronization of the activity of certain neurons plays a defining role [MAL 81], and the simultaneous recording of multiple neurons strongly facilitates or is perhaps even necessary for the experimental study of this kind of model [DON 08];

3) multiple examples such as that of the motor system [GEO 86] show that, even without synchronization, *groups of neurons* are required to properly represent a stimulus or an action such as a motor command; even though it is sometimes possible to study these phenomena through successive recordings of unique neurons (as was performed in [GEO 86]), simultaneous recordings make this task a lot easier (which returns to the first point outlined above).

8.1.2. *How can neuronal populations be recorded?*

There are currently three methods for recording neuronal populations. Multiple extracellular recordings [BUZ 04] are the most commonly used technique. The subject of this chapter is the analysis of data obtained using this technique. Recordings with potential-sensitive probes [ZEC 89, HOM 09] are used on brain slices and ganglia of invertebrates. Finally, recordings using calcium-based fluorescence [HOM 09] are often presented in the literature, but their resolution in time is insufficient for the study of questions of synchronization (see Figure 4 in Chapter 4 of [CAN 10]).

1 The technique of EEG, extensively discussed throughout this book, also provides simultaneous recordings of multiple neurons, but without the resolution of individual neurons.

2 See: http://en.wikipedia.org/wiki/Binding_problem.

8.1.3. *The properties of extracellular data and the necessity of spike sorting*

Figure 8.1 shows 1 s of recording at the four sites of a *tetrode* [GRA 95]. This recording was performed in the first olfactory relay, the *antennal lobe*, of an insect: the locust *Schistocerca americana*. These readings will serve as a running example throughout this chapter. Before being converted into a numerical format (at a sample rate of 15 kHz), the data were filtered between 300 Hz and 5 kHz; the full details of the recordings are given in [POU 02]. The reader should note that the 300 Hz high-pass filter will have removed most of the *local field potentials* that arise from postsynaptic activity. To keep this chapter brief, we will not discuss the analysis of this type of signal; they are identical to signals obtained by intracranial EEG[3].

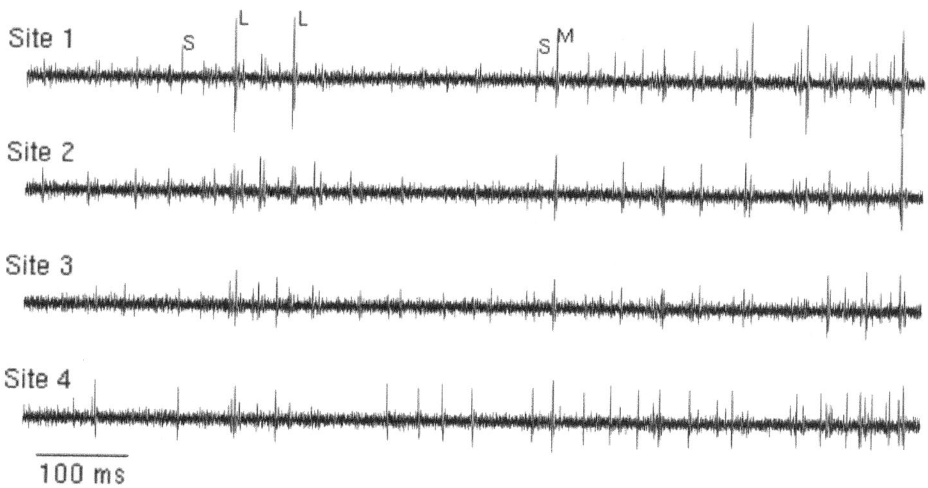

Figure 8.1. *One second of data recorded at the four sites (electrodes) of a tetrode. The data were filtered between 300 Hz and 5 kHz before being converted into a numerical format. The sample rate is 15 kHz. These readings were taken in the antennal lobe of a locust* Schistocerca americana. *Examples of action potentials at the first site are marked by the letters S, M and L, with small, medium, and large amplitudes, respectively*

3 See: https://en.wikipedia.org/wiki/Electrocorticography.

The spikes visible at each of the sites of the recording, some examples of which have been marked by the letters S (small), M (medium) and L (large) in Figure 8.1, are of particular interest to us. These spikes are generated by action potentials emitted by multiple neurons [PLO 07]. Based on the "all-or-nothing" property of action potentials *in axons*, we can conclude, at least provisionally, that multiple neurons are present [ADR 22]. Given these type of data, we start by asking the following two interrelated questions:

1) How many neurons contributed to the recording?

2) Which neuron was the originator of each of the visible spikes?

Spike sorting is the stage of data processing that attempts to answer these questions; we will study it in this chapter. We will return to the question of the origin of the signal and a justification of the use of a tetrode, which consists of multiple electrodes in close proximity of each other. The reader should, however, note at this point that the spikes marked with S appear to be associated with signals of similar amplitude on the fourth site, even smaller amplitude on the third site and zero apparent amplitude on the second site. In contrast, the amplitude of the signals associated with spikes marked with M (medium) appears to be constant across all sites. We will see that the distance between the source of the current, i.e. the neuron, and the electrode is the principal determining factor of the recorded signal; given a fixed source (neuron), the *amplitude ratios* are thus functions of the distance ratios between the source and the recording sites: *they depend on the position of the source*. Therefore, different ratios correspond to different sources (neurons).

8.2. The origin of the signal and its consequences

8.2.1. *Relationship between current and potential in a homogeneous medium*

The equation that governs the relationship between the current emitted by a point source[4] of intensity I_0 (in amperes) and electrostatic potential Φ_e (in

4 Point sources of current are conceptual constructs, in which they does not exist physically, but provide a relevant approximation when a "neurite element" such as an axon segment measuring 0.5 μm or even a soma measuring 15 μm is recorded by an electrode placed at a distance of 100 μm.

volts) observed at a distance r (assuming the potential at infinity is zero) is given by:

$$\Phi_e = \frac{1}{4\pi\sigma_e}\frac{I_0}{r}, \qquad [8.1]$$

where σ_e is the conductivity of the extracellular medium (in Siemens per meter) assumed to be uniform. We shall neglect certain capacitive properties of the extracellular medium [BED 04], but the model developed here, nonetheless, provides an excellent initial approximation [LIN 14]. It is clear that in order to apply equation [8.1], we must first have some method of measuring or estimating the current. The most common approach is to assume a realistic neuron morphology by building in various conductances distributed in a non-homogeneous manner in the membrane, and then numerically solving the cable equation. Solving this equation yields the densities of the various currents. Their sum $i_m(x)$ (where x denotes the lengthwise position within the cable) is then used in a differential version of equation [8.1] to obtain a formula of the following type:

$$\Phi_e = \frac{1}{4\pi\sigma_e}\int_N \frac{i_m(x)}{r(x)}dx, \qquad [8.2]$$

where the integral is taken along the skeleton of the neuron[5], referred to by the label N. If the value of the membrane potential is known "at each point" along the neuron, then by an elementary application of Ohm's law and the law of conservation of charge the desired current density may also be obtained. Warning to the reader – the next section requires some mental gymnastics. The membrane potential is usually represented at a fixed position as a function of time, for example in the presence of an action potential. In the next section, we shall fix the time and vary the position. Note that in the case of an action potential propagating at constant speed without deformation, the second representation may be easily deduced from the first.

5 By skeleton, we mean the true morphology of the neuron after reducing the diameter of each neurite to 0; it is effectively a one-dimensional object with branches embedded in three-dimensional space.

8.2.2. *Relationship between the derivatives of the membrane potential and the transmembrane current*

As described by Rall [RAL 77, pp. 64–65] and Plonsey and Barr [PLO 07, Chapter 8], let us consider a "small segment of neurite" of radius a and length Δx. If the intracellular potential[6] $\Phi_i(x,t)$ at one of the ends of the segment is not equal to the potential $\Phi_i(x+\Delta x,t)$, then Ohm's law states that there is an axial current of intensity:

$$\begin{aligned} I_i(x,t) &= -\pi a^2 \sigma_i \frac{\Phi_i(x+\Delta x,t) - \Phi_i(x,t)}{\Delta x} \\ &\approx -\pi a^2 \sigma_i \frac{\partial \Phi_i(x,t)}{\partial x}, \end{aligned} \quad [8.3]$$

where σ_i is the intracellular conductivity, and where currents are taken to be positive in the direction of increasing x. Now, if the current $I_i(x,t)$ entering the segment is not equal to the current $I_i(x+\Delta x,t)$ exiting the segment, the law of conservation of charge states that the difference must have passed through the membrane; since the density of the transmembrane current $i_m(x,t)$ is positive for outward-flowing current, we obtain:

$$\begin{aligned} I_i(x+\Delta x,t) - I_i(x,t) = -\Delta x\, i_m(x,t) \quad \text{giving} \\ \frac{\partial I_i(x,t)}{\partial x} = -i_m(x,t), \end{aligned} \quad [8.4]$$

from which we obtain, after combining with equation [8.4]:

$$i_m(x,t) = \pi a^2 \sigma_i \frac{\partial^2 \Phi_i(x,t)}{\partial x^2}. \quad [8.5]$$

Given that the gradients of the observed potentials immediately outside the membrane are much lower than those inside (because the resistance between two external points is much lower than the resistance between two interior points), equation [8.5] becomes:

$$i_m(x,t) = \pi a^2 \sigma_i \frac{\partial^2 V_m(x,t)}{\partial x^2}.$$

6 We explicitly include time, even though initially time is assumed to be fixed.

where V_m is the transmembrane potential, and equation [8.2] may be rewritten as:

$$\Phi_e = \frac{a^2 \sigma_i}{4\sigma_e} \int_L \frac{1}{r(x)} \frac{\partial^2 V_m(x,t)}{\partial x^2} \, dx \,. \qquad [8.6]$$

Let us now consider the case of two long neurites with identical properties[7] except for their radius, both capable of propagating an action potential. In the fourth volume of their monumental series, Hodgkin and Huxley solved the wave equation (equation 30 in [HOD 52]). that is to say the ordinary differential equation satisfied by an action potential propagating at constant speed[8]. They also showed that the time scale of the membrane potential does not depend on the radius of the axon, and the speed of propagation θ of the action potential satisfies:

$$\frac{\theta^2}{a\,\sigma_i} = K \,, \qquad [8.7]$$

where K is a constant. Using dimensional analysis, Goldstein and Rall [GOL 74] also showed that the size of the action potential in space is proportional to the square root of the radius – it grows at the same rate as the speed – which implies that, considered as a function of time at any given point of the axon, the action potential does not depend on the radius. These results only hold for *non-myelinated* fibers. The effect of the radius on the spatial profile of the action potential and on its second derivative is shown in the upper section in Figure 8.2, which considers axons of radius 1 μm (left) and 2 μm (right)[9]. We can clearly see that the spatial breadth increases with the diameter, and that the second derivative (necessarily) decreases twice as rapidly with the diameter. The term $1/r(x)$ from equation [8.6] is shown in black (for an electrode situated 50 μm from the center of the axon). Since an

7 That is, with identical conductances (types and values) and identical plasmic resistivity.

8 As a historical aside, Hodgkin and Huxley did not solve the system of equations that now bears their names, which involves an equation with partial derivatives, but they did solve – with a mechanical calculator – the simpler system satisfied by a wave propagating without deformation along an axon, i.e. an action potential.

9 This results come from numerical solutions of the equations of Hodgkin and Huxley obtained using the "classical" parameters specified by them (Detorakis and Pouzat, manuscript in preparation).

analytical solution of equation [8.6] is not possible, numerical solutions for axons with radii between 10^{-1} and 20 μm – the domain of observed values in non-myelinated cortical fibers – are summarized in the graphs at the bottom of the figure. The bottom-left graph of Figure 8.2 shows that the extracellular potential grows faster than the radius to power 1.8; the right-hand graph shows that the extracellular potentially decreases independently of the radius at least as rapidly as one over the square of the distance between the electrode and the axon.

Thus, if an axon of diameter 0.5 μm is connected to a soma of diameter 15 μm, the extracellular signal will be dominated by whatever happens inside the soma (Figure 16 in [FAT 57]); otherwise, the axon can be ignored without affecting the value of the extracellular potential. The action potentials recorded by extracellular electrodes will therefore reflect the events that unfold inside the soma and the apical dendrite, if active. This also explains why it is considerably easier to record pyramidal cells (large neurons) than interneurons (small neurons) [GRO 70]. Additionally, if the action potentials of the soma are not identical to those of the axon, as was demonstrated to be the case in experiments[10] for both invertebrates (Figure 13 in [EYZ 55]) and vertebrates (Figure 5A in [WIL 99]), then the relationship between the action potentials recorded by extracellular electrodes and the effective emissions of the neuron is probably not uniquely characterized – the relation: `one somatic action potential` = `one action potential in the axon` probably does not always hold. We must also consider that propagation may fail at the branching points of the axon (Figure 7 in [ANT 00]) and the possibility of "reflection" of action potentials (Figure 5 in [ANT 00]); two phenomena that may only be properly accounted for with recordings "at all points" in the axon[11]. The conclusion of this short section is that critical thinking remains valuable in the analysis of sequences of action potentials obtained by extracellular recordings (and indeed by intracellular recordings) of the somatic system.

10 These experiments show that it is possible to have a very small action potential at a somatic level – these somatic recordings are intracellular, which means that these action potentials are very likely indistinguishable from noise in the context of extracellular recordings – together with a perfectly typical action potential in the axon during high-frequency discharges, or *bursts*.

11 Recordings may be obtained for one unique neuron with membrane potential-sensitive probes.

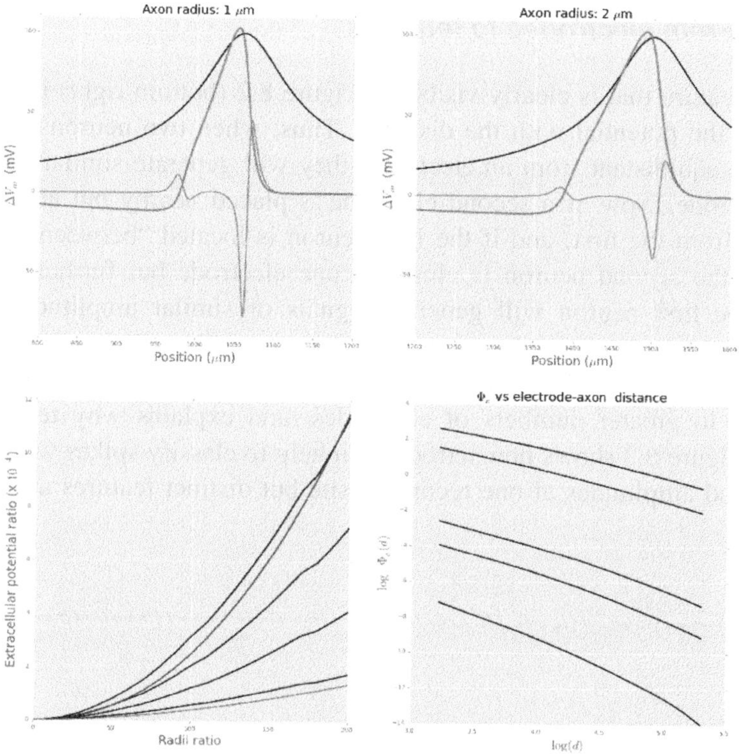

Figure 8.2. *Top: the membrane potential (in red) – expressed as the deviation relative to the rest value – of an action potential of two axons differing only in their radius, 1 μm on the left and 2 μm on the right. In blue, the second derivatives of V_m with respect to the position; the peak value on the left is 0.46 $\mathrm{mV}/\mu\mathrm{m}^2$, and is equal to half of this on the right. In black, the curve shows the term $1/r(x)$ from equation [8.6] for an electrode situated at 50 $\mu\mathrm{m}$ from the center of the axon whose position along the axon is given by the minimum point of the second derivative of V_m; its peak value is 2×10^{-2} $\mu\mathrm{m}^{-1}$. The integrand of equation [8.6] is the product of the blue curves with the black curves. Bottom, a summary of numerical solutions for axons with radii between 10^{-1} and 20 $\mu\mathrm{m}$. On the left, the (minimum values of) Φ_e over the value of Φ_e at a radius of 10^{-1} $\mu\mathrm{m}$ as a function of the ratio of the radii. The y-values of the red curve are the x-values to power 1.8, and those of the blue curve are to power 2.2. The various black curves show, from bottom to top, electrodes placed at 25, 50, 75 and 100 $\mu\mathrm{m}$ from the center of the axon. On the right, diagrams showing the evolution of (the negative of the minimum value of) Φ_e for a given axon radius as a function of the distance d between the electrode and the center of the axon. The radii of the axons are from bottom to top: 10^{-1}, 5×10^{-1}, 1, 10 and 20 $\mu\mathrm{m}$. For a color version of this figure, see www.iste.co.uk/clerc/interfaces1.zip*

8.2.3. "From electrodes to tetrodes"

One feature that is clearly visible in Figure 8.2 (bottom right) is the $1/r^2$ decay of the potential with the distance. Thus, when two neurons "of same type" are equidistant from an electrode, they will generate similar signals at that electrode. Now, if a second electrode is placed nearby but at a distinct location from the first, and if the first neuron is located "between the two", whereas the second neuron is closer to one electrode but further from the other; the first neuron will generate signals of similar amplitude at both electrodes, and the second neuron will generate a "large" signal at the first electrode and a "small" signal at the second electrode. This reasoning may be extended to greater numbers of electrodes, and explains why tetrodes are used[12]. Figure 8.3 shows how tetrodes can help to classify spikes with similar shapes and amplitudes at one recording site but distinct features at the other sites.

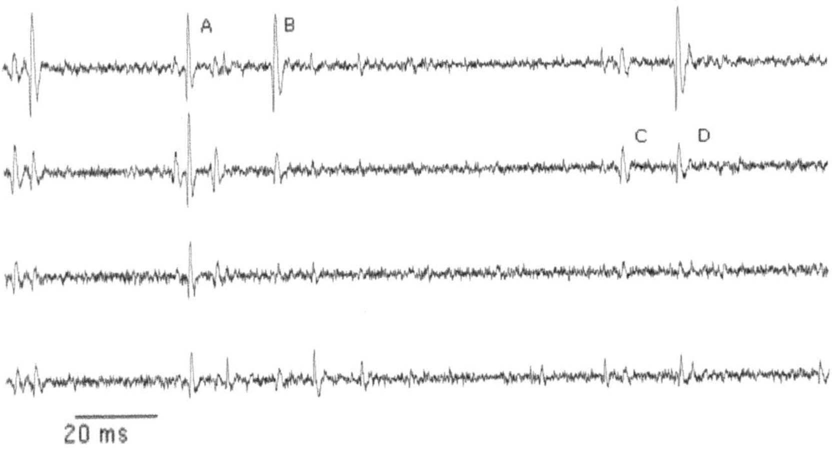

Figure 8.3. *Two hundred millisecond of data from the recording shown in Figure 8.1. A and B mark two action potentials with amplitudes and shapes that are similar at site 1, but very different at the three other sites. Similarly, C and D mark two action potentials that are similar at site 2, but different at the other sites*

12 However, tetrodes are not always useful; for example, they serve no purpose in the antennal lobe of the cockroach *Periplaneta americana*, while in the present example, in the antennal lobe of the locust *Schistocerca americana*, they are essentially indispensable.

8.3. Spike sorting: a chronological presentation

We continue with a "chronological" presentation of the principal methods of spike sorting. This approach is not particularly synthetic, but in our opinion it introduces the various relevant problems and solutions into a concrete setting with minimal formalism. The figures in each section are drawn from actual data. A few simplifications were made, such as the use of a single recording site when four sites were available. The only differences between these examples and what is done "in practice" are technical in nature, the *ideas are the same*, and it is the ideas that are important. The data and a full step-by-step description of its analysis with the software packages R[13] and Python[14] are available on the author's website[15].

8.3.1. *Naked eye sorting*

When the "all-or-nothing" property of action potentials in the axon was first established [ADR 22], the only tools available to neurophysiologists were recordings on paper (the oscilloscope had not yet been invented), and they had to laboriously carry out sorting with the naked eye based on amplitude (Figure 4 in [HAR 32]), in a somewhat similar fashion to the way that we analyzed the first site of Figure 8.1.

8.3.2. *Window discriminator (1963)*

Once magnetic tape recording systems had become commonplace in physiology labs, the quantity of data to be processed increased significantly, with the immediate result of inspiring certain researchers to automate the processes that they had previously been performing by hand [POG 63]. The first innovation was to construct dedicated electronic circuits. Samples were classified by the peak amplitude of their events[16] as illustrated in Figure 8.4.

---

13 http://www.r-project.org/.

14 https://www.python.org/.

15 http://xtof.perso.math.cnrs.fr/sorting.html, at the bottom of the page.

16 This method is still used today in some labs, especially those that need to perform sorting in real time. It is also used systematically for the audio outputs of amplifiers. Experimental researchers listen to the output of one of the electrodes when inserting them into the tissue; and the electronic circuit between the amplifier output and the speaker removes all amplitudes

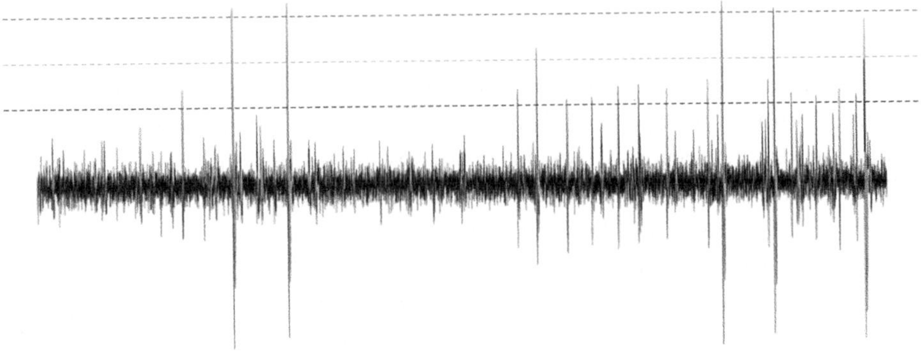

Figure 8.4. *Principle of the window discriminator. Three zones corresponding to three classes of action potentials and thus three "neurons" are defined here: small action potentials are those whose peak value is located between the black and orange dotted lines; medium action potentials have a peak value between the orange and blue lines, and large action potentials have a peak value higher than the blue line. The data were taken from the first site in Figure 8.1. For a color version of this figure, see www.iste.co.uk/clerc/interfaces1.zip*

8.3.3. *Template matching (1964)*

Physiologists soon realized that action potentials originating from two different neurons could have the same peak amplitude but different shapes (Figure 8.5(a)). This led to the introduction of a two-step method [GER 64]:

1) Two events with the same class of shape or template were identified with the naked eye, and for each pattern, a dozen or so events were averaged. These averages would subsequently serve as template estimators.

2) Each event was compared to each template by subtracting the template from the event and calculating the sum of the squares of the components of the difference vector, i.e. the *residual* vector (Figure 8.5(b)). The event is then assigned to the closest template, that is to say the template with the smallest residual vector.

---

below a given threshold and saturates all amplitudes above a second threshold. Since large spikes will be above the second threshold for longer than short spikes, the amplitude of each event is encoded into the duration of the sounds.

In statistics, an *estimator*[17] is a function of the data that provides an estimate of a parameter. Here, the parameters are the templates, or more concretely an ordered sequence or *vector* of 45 amplitudes (in the original article [GER 64], these vectors were defined by 32 amplitudes). Estimators are functions of the data, so the value of an estimator changes as the data changes; formally, they are *random variables*.

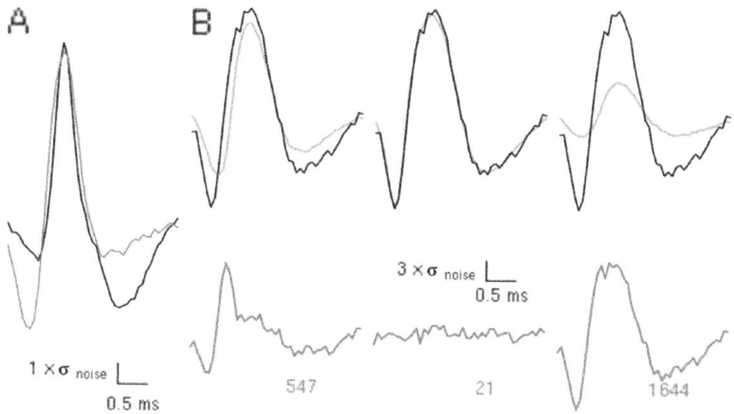

Figure 8.5. *Principle of template matching A), the templates of neurons 6 (in red) and 7 (in black) (at the fourth site). B) top, the same event (in black) and three of the 10 templates (in gray) at the first site; below, the templates were subtracted from the event, and the corresponding sum of the squares of the residues is shown in red. The event clearly matches the second template. For a color version of this figure, see www.iste.co.uk/clerc/interfaces1.zip*

8.3.4. *Dimension reduction and clustering (1965)*

Faced with the problem of low computer memory availability in physiology labs, Simon [SIM 65] had the idea that we should avoid working with the full sequence of amplitudes as required by the technique of template matching, instead restricting attention to the amplitudes measured at two carefully chosen points in time of the event (Figure 8.6(a)). These points in time were selected by observing the superimposed events together, so that the amplitudes of the different categories of spike would be distinguished as clearly as possible. This technique was able to reduce the number of

17 See: https://en.wikipedia.org/wiki/Estimator.

parameters necessary for characterizing an event by 30 or more; graphically, we go from A to B in Figure 8.6. *Disjoint* domains in the plane representing the data were constructed by the user "by hand", each domain corresponding to one neuron. Events were then classified according to the domain to which they belonged (Figure 8.6(c)). Once this classification had been performed on the reduced-dimension space, it is still possible, and perhaps even advisable, to return to the initial representation in order to review the results (Figure 8.6(d)). In today's terminology, we would say that we *reduced the dimension*[18] by passing from A to B in Figure 8.6. This process of defining domains is an example of what is now known as *clustering*. These two very important aspects of high-dimensional data analysis – of which spike sorting is an example – are described in a manner that is both general and very pedagogical in the book by Hastie *et al.* [HAS 09]. The two spaces between which we have been moving, the 45-dimensional space (Figures 8.6(a) and 8.6(d)) and the 2-dimensional space (Figures 8.6(b) and 8.6(c)), are called *sample spaces*[19] by statisticians [BRE 09].

8.3.5. *Principal component analysis (1968)*

The fact that the user must choose two points at which to compare the amplitude as coordinates for the reduced space is inconvenient in the previous method. Physiologists therefore kept looking for alternative methods, more automatic and more efficient. Principal components analysis [GLA 68] was the first such alternative to be proposed, and today remains the most widely employed technique. Principal components analysis finds the subspace of desired dimension that reproduces the largest possible fraction of the variance of the sample – here, the term sample is used in the statistical sense: a set of observations/individuals randomly selected from a population. We will not discuss this method further here [GLA 76, HAS 09], but we will mention the fact that typical applications involve the intermediate step of singular value decomposition[20] of the covariance matrix of data, which we shall briefly explain in the following note.

18 See the section on dimension reduction at: https://en.wikipedia.org/wiki/Dimensionality_reduction.

19 The sample space is the set of all potentially observable events in an experiment. The first step of probabilistic modeling is to define this space.

20 See: https://en.wikipedia.org/wiki/Singular_value_decomposition.

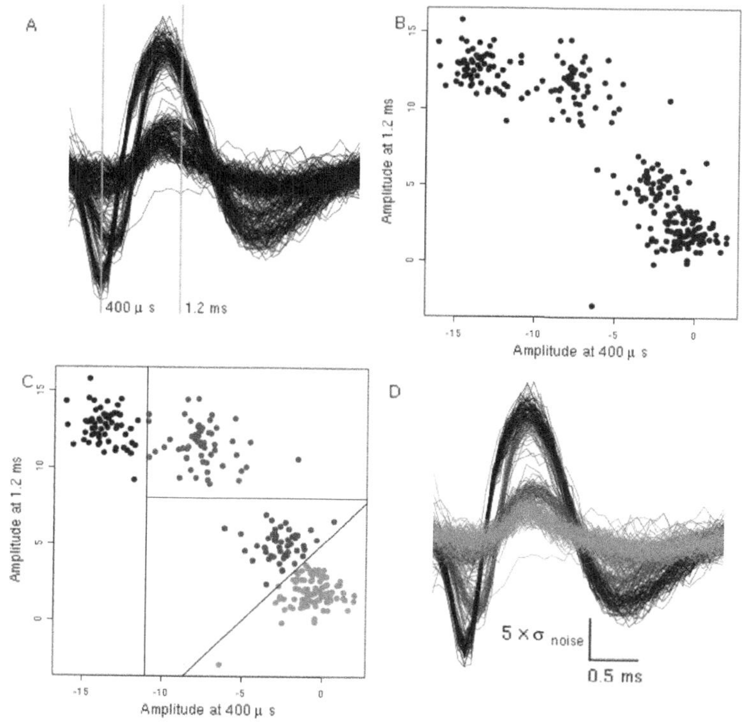

Figure 8.6. *Principle of dimension reduction and clustering. A) 267 spikes recorded at the first site. The vertical gray lines at 400 μs and 1.2 ms indicate the two chosen points in time. B) Amplitude at 1.2 ms as a function of the amplitude at 400 μs. C) Same as (B), except the boundaries of a partition have been defined manually; the spikes are colored according to the class that contains them. D) Same as (A), except the spikes are colored according to the class that contains their projections. For a color version of this figure, see www.iste.co.uk/clerc/interfaces1.zip*

NOTE.– The covariance matrix is constructed from the matrix D of data, whose rows are the events. In Figure 8.6, D is a matrix with 267 rows – there are 267 events – and 45 columns – each event e_i is specified by an ordered set of 45 amplitudes $(e_{i,1}, \ldots, e_{i,45})_{i=1,\ldots,267}$. Each of the action potentials in Figure 8.6(a) corresponds to one row of the matrix D. The covariance matrix is obtained by subtracting the average row $(\overline{e}_j)_{j=1,\ldots,45}$ where $\overline{e}_j = \sum_{i=1}^{267} e_{i,j}/267$ from each of the rows $(e_{i,1}, \ldots, e_{i,45})_{i=1,\ldots,267}$, which yields the matrix M whose entries are given by $M_{i,j} = e_{i,j} - \overline{e}_j$. The entries of the covariance matrix V are then given by $V_{i,j} = \sum_{k=1}^{267} M_{i,k}\, M_{k,j}/267$, or, written as matrix multiplication: $V = M^T M/267$, where M^T is the transpose of M (See https://en.wikipedia.org/wiki/Covariance). The result of principal component analysis on the set of data in the previous section is shown in Figure 8.7. Figure 8.7(a) shows both the average event (in black) and the average event plus the first (in red) and second (in blue) principal component multiplied by 10. We see that events whose projections onto the first component have high values differ from the average in amplitude but not in shape, and that events whose projections onto the second component have high values differ in shape but not in amplitude. The reader should note (Figure 8.7(b)) that the multiplicative factor of 10 is the same order of magnitude as the observed values. Figure 8.7(b) corresponds to Figures 8.6(b) and 8.6(c) and shows that it is easy to define domains by reducing the dimension along the principal components. Nevertheless, we should note that performing principal component analysis requires a certain amount of (computer) memory, the absence of which was precisely what originally motivated Simon [SIM 65] to introduce the idea of dimension reduction. These memory constraints have now long since disappeared.

8.3.6. *Resolving superposition (1972)*

Since the deviations of the extracellular potentials of action potentials are of the order of the millisecond, we can expect to observe instances of superposition[21] similar to those shown in Figure 8.8(a) whenever sufficient neurons are registered by the recording[22]. This phenomenon was characterized in the early 1970s, and solutions based on “manual” template matching were suggested [PRO 72]. Clearly, as shown in Figure 8.8, resolving superposition requires the templates to have been estimated: superposition cannot be resolved simply by considering the projection of the data onto a subspace like in the previous two sections. Today, the most

21 In the literature, the terms “collision” and “interference” are also used to describe this phenomenon.

22 If ν is the average discharge frequency of K neurons in a recording, and Δ is the typical duration of an action potential and if we assume that neuron discharges may be modeled sufficiently accurately by a Poisson distribution, then the probability of there being zero action potentials within a window of duration Δ is $\exp -K\nu\Delta$, the probability of there being exactly one is $K\nu\Delta \exp -K\nu\Delta$ and the probability of there being at least two is $1 - (1 + K\nu\Delta) \exp -K\nu\Delta$; the frequency of superposition among windows containing at least one event is the ratio of these last two values.

commonly used methods of spike sorting are derived from methods that combine dimension reduction and clustering; since these techniques are not capable of resolving superposition, it appears that the majority of published results that rely on sorting simply ignore superposition.

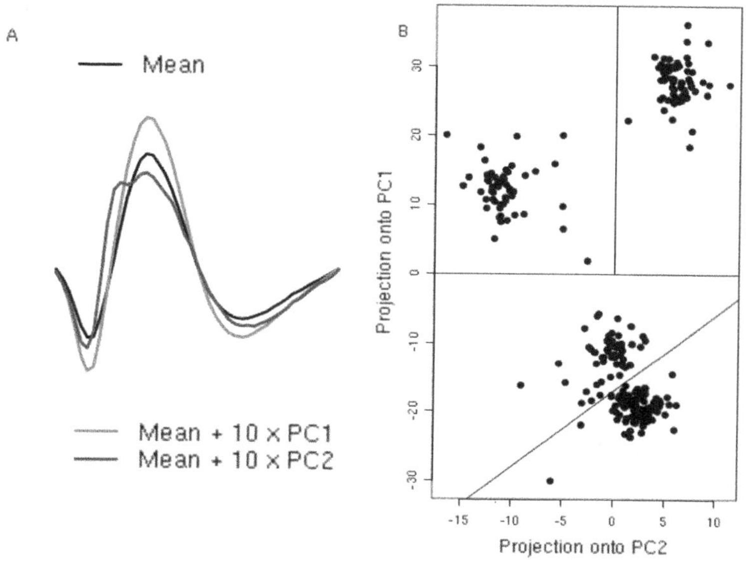

Figure 8.7. *Principal component analysis. A) In black, the average of the 267 action potentials recorded at the first site; in red, the same average plus 10 times the first principal component; in blue, the same average plus 10 times the second principal component. B) Two hundred sixty-seven events projected onto the plane defined by the first two principal components. Domains that yield the same classification as in Figure 8.6 have been added by hand. For a color version of this figure, see www.iste.co.uk/clerc/interfaces1.zip*

8.3.7. *Dynamic amplitude profiles of action potentials (1973)*

Until the early 1970s, recordings of isolated axons or nerves were very common, especially in invertebrates. In this type of recording, the "all-or-nothing" property of action potentials effectively holds, even during high-frequency firing. But as cortical recordings in vertebrates became more commonplace, a new problem particular to these subjects soon presented itself: the dynamic character of the amplitude profiles (and sometimes the shape) of the action potentials emitted by a neuron during high-frequency or *burst* discharges, as shown by Figure 8.9(a). The solution suggested by Calvin [CAL 73] requires manually processing the data, and relies on a "relatively" stable combination of amplitude reduction and interspike intervals during bursts (Figure 8.9(c)). Note how these amplitude dynamics introduce additional obstacles for spike sorting (Figure 8.9(b)).

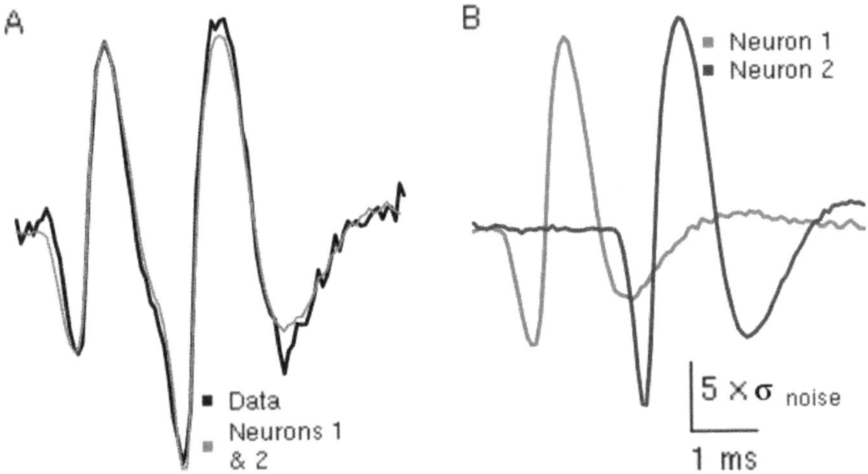

Figure 8.8. *Resolving superposition. A) In bold black, an event that does not correspond to any of the templates/neurons; in gray (thinner) the sum of the two patterns – associated with neurons 1 and 2 of our classification – shown in B). The scales of the two graphs are identical. For a color version of this figure, see www.iste.co.uk/clerc/interfaces1.zip*

8.3.8. *Optimal filters (1975)*

Attempts to perform multiple recordings along the nerve of a marine invertebrate led Roberts and Hartline [ROB 75] to suggest a method capable of automatically decomposing instances of superposition. Their method may be viewed as an extension of the template matching method; the idea is to construct one filter per neuron such that the filter is maximal when an action potential emitted from the neuron *for which the filter was constructed* is present in the data, and minimal or zero when noise or emissions from another neuron are running through the signal. The filters are linear, so that if action potentials of two or more neurons are present with a small offset in time, such as in Figure 8.8(a), the output of the two filters should display spikes with the same offset in time. Similarly to the technique of template matching, the method assumes that we have previously estimated the characteristic shapes/patterns associated with each neuron, at each recording site if multiple sites are in use. The construction of optimal filters is slightly too complicated to be fully explained here ([ROB 79] has all of the details), but we will illustrate the idea with the example of *matched filters*[23]. The characteristic shapes of the neurons obtained at each of the sites (three templates corresponding to three different neurons are shown in gray in the upper section of Figure 8.5(b)) are represented by a set of vectors, with one vector per recording site. Each vector has the same number of elements, corresponding to the number of sample points in the template – this method works best with lengthy templates that start at zero and return to zero, whereas in general the method of template matching works well even with shorter templates – in the case shown in Figure 8.10, for the second neuron, we have 130 points per template at sites 1 and 4:

$$\mathbf{m}_2 = \begin{pmatrix} m_{2,1} \\ m_{2,4} \end{pmatrix} = \begin{pmatrix} m_{2,1,1}, \ldots, m_{2,1,130} \\ m_{2,4,1}, \ldots, m_{2,4,130} \end{pmatrix}$$

To construct a *matched filter* from these two vectors, we begin by subtracting from each $m_{2,i,j}$ the average at the corresponding site: $m_{2,i,\bullet} = \sum_{j=1}^{130} m_{2,i,j}/130$ to obtain $f_{2,i,j} = m_{2,i,j} - m_{2,i,\bullet}$, and then we normalize so that the scalar product of the filter $\mathbf{f}_2$ with the original template ($\sum_{i\in\{1,4\}} \sum_{j=1}^{130} m_{2,i,j} f_{2,i,j}$) is equal to one. In the suboptimal case of a *matched filter*, the filters are therefore just normalized versions of the templates. If we write the data to which the filter will be applied in the following form:

$$\begin{pmatrix} \ldots, d_{1,k-2}, d_{1,k-1}, d_{1,k}, d_{1,k+1}, d_{1,k+2}, \ldots \\ \ldots, d_{4,k-2}, d_{4,k-1}, d_{4,k}, d_{4,k+1}, d_{4,k+2}, \ldots \end{pmatrix}$$

then the output $F_{2,k}$ of filter 2 at "time" k is given by the expression:

$$F_{2,k} = \sum_{i\in\{1,4\}} \sum_{j=1}^{130} f_{2,i,j} d_{i,k+j-J} \, ,$$

23 See: http://en.wikipedia.org/wiki/Matched_filter.

where J is the position of the peak within the template m_2. The implementation of this method, and the way that it can automatically resolve superposition, are demonstrated in Figure 8.10. *Matched filters* are suboptimal because their "interference", that is to say the output of the filter for templates that it was not designed to match, has not been optimized. The secondary peak in the output of filter 2 shown in Figure 8.10(B) is one such example of interference. The method of filter construction presented in [ROB 79] reduces this problem significantly; nevertheless, if too many of the characteristic shapes are too similar it will not be possible to fully eliminate the interference.

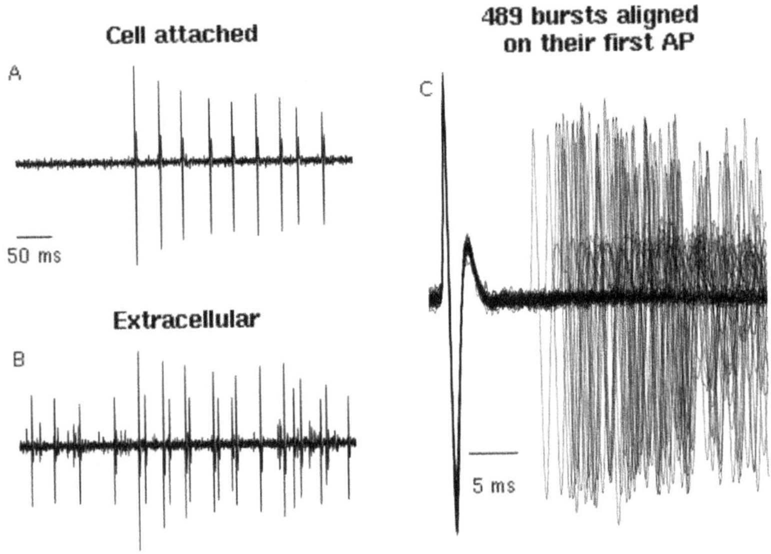

Figure 8.9. *Dynamic amplitude profiles of action potentials A), example of a burst, somatic recording in a "cell-attached" arrangement of a Purkinje cell in a slice of cerebral cortex of a rat. Note how the amplitude of the action potentials is diminished during the burst. B) Simultaneous extracellular recording (recordings taken by Matthieu Delescluse). Three neurons, including the one recorded in cell-attached mode, register on this recording. Notice how the action potentials of a tonically active neuron have a similar amplitude to the action potentials of the reference neuron (also recorded in a cell-attached arrangement) at the end of the burst. C) Four hundred eighty-nine recorded bursts over the course of 1 min aligned by their first action potential. The details of the recordings and data processing steps are given in [DEL 06]*

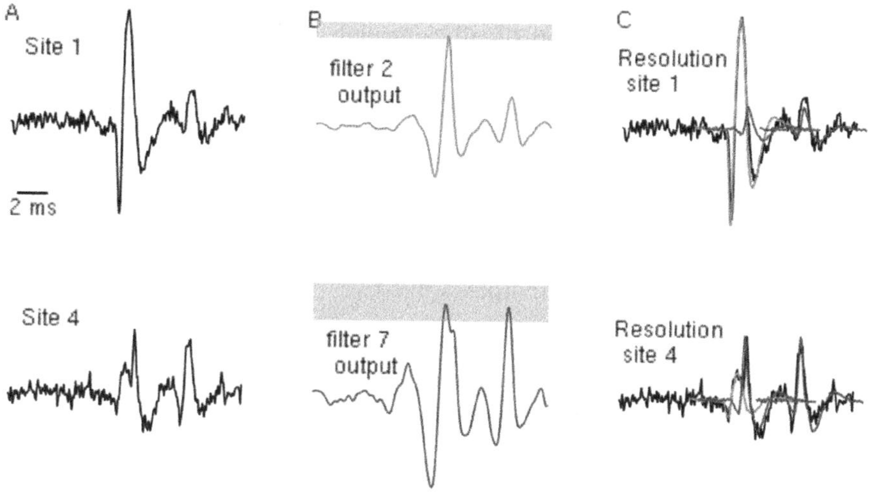

Figure 8.10. *Matched filters A), a subset of the data from the grasshopper example, at two of the four sites of recording. B) Corresponding output of the matched filters constructed from the templates of neurons 2 and 7. The gray bands are 99% confidence bands obtained by superimposing "noise events" – segments of raw data between the two detected action potentials – and the template of each of the two neurons before applying the filter. C) The resolved signals obtained from the filter outputs. For a color version of this figure, see www.iste.co.uk/clerc/interfaces1.zip*

8.3.9. *Stereotrodes and amplitude ratios (1983)*

The most direct and possibly still the most effective method of sorting spikes with dynamic amplitude profiles was suggested by McNaughton *et al.* [MCN 83]. It is perhaps not so much a method of analysis, but rather a recording technique: stereotrodes (two recording sites in close proximity, as suggested in the original article) or tetrodes (four sites in close proximity [GRA 95]). The motivation for this method is presented in perfect clarity in the penultimate paragraph of their introduction:

> "The method described in the present report is based on the fact that the size of the extracellular action potential varies inversely with the distance of the recording electrode from the current generator. In theory, a closely spaced tetrahedral array of recording electrodes with tips sufficiently close together to record

> signals from overlapping populations of neurons should permit the unique identification of all neuronal spikes that exceed the noise level. This is so since each cell would generate a unique point in the three-dimensional data space whose axes are defined by the spike height ratios of channels 1 and 2, 2 and 3, and 3 and 4. Note, that since the discrimination is based on amplitude ratios, the problem of intrinsic variation in spike amplitude such as occurs during the complex spike burst of hippocampal pyramidal cells is, in principle, solved".

The data recorded in slices of the cerebral cortex of a rat, which we previously used in Figure 8.9, will once again serve to illustrate the principle of amplitude ratios. Figure 8.11 shows 200 ms of data recorded at the two sites (separated by 50 μm) of a stereotrode. The action potentials from the "reference" cell in this last figure are marked with vertical gray dotted lines. Action potentials from a different cell that fires "in pairs" (and sometimes in triplets) with strongly characteristic amplitude dynamics are also marked with vertical gray lines.

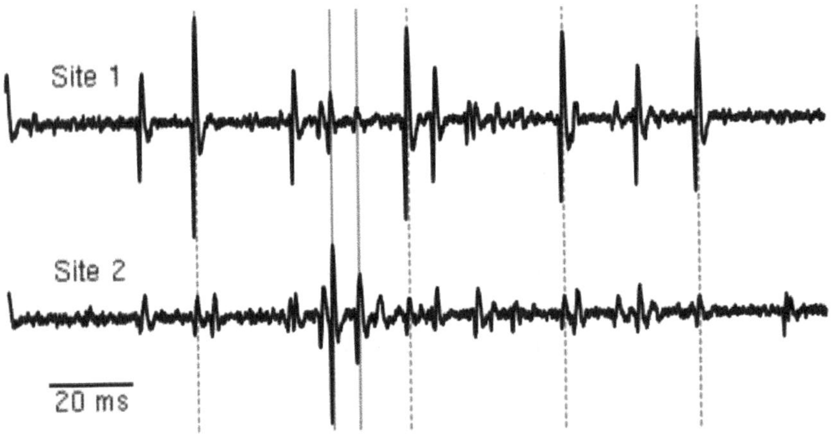

Figure 8.11. *Data from a stereotrode 200 ms of extracellular data recorded along the surface of the cell bodies of Purkinje cells – cerebral cortex slice of a young rat, same dataset as in Figures 8.9. The vertical gray lines mark two action potentials of a cell firing in pairs (with a strongly dynamic amplitude profile). The vertical gray dotted lines show the action potentials of the burst-firing cell if Figure 8.9*

After detecting the action potentials by identifying the local maxima above a certain threshold, the peak amplitudes of each spike are obtained, and each action potential is represented in Figure 8.12 (left) as a point on a plane (sample space) whose axes are given by the peak amplitude at the second site (horizontal) and the peak amplitude at the first site (vertical). Each point is assigned an angle by calculating the arctangent of the amplitude ratio. Calculating the amplitude ratio is always a somewhat sensitive operation, because dividing two noisy values increases the error. In order to avoid excessively large errors, we performed regression on the amplitudes near the peak at site 1 (5 points on each side of the peak) as a function of the corresponding amplitudes at site 2, neglecting the constant term. The estimated density of the angles is shown in Figure 8.12 (right). Thus, we obtain well-defined peaks, which may be used to define angular domains corresponding to domains of amplitude ratios.

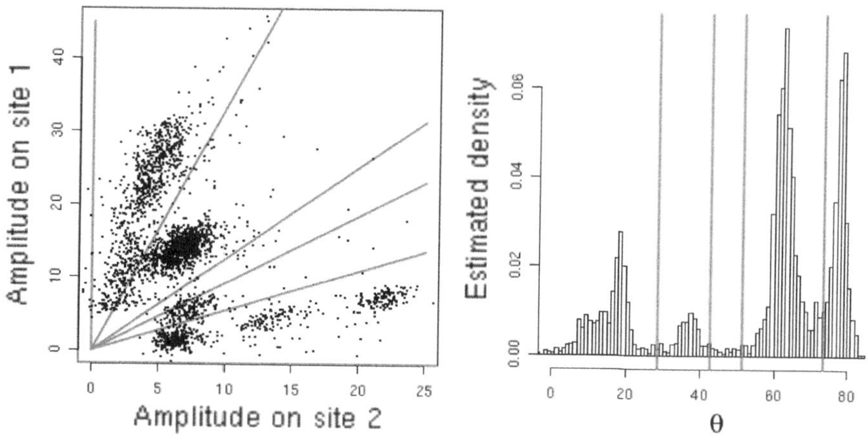

Figure 8.12. *Amplitude ratios. On the left, the (peak) amplitude at site 1 as a function of the (peak) amplitude at site 2 for the action potentials detected in the dataset in Figure 8.11. The units of the axes are standard deviations of the noise. The gray lines correspond to the angular domains defined in the right-hand section. On the right, the distribution of the θ angles estimated using the tangent obtained by regressing the 10 amplitude values in the neighborhood of the peak at site one as a function of the 10 amplitude values in the neighborhood of the peak at site 2 (neglecting the constant term). The vertical gray lines were placed "with the naked eye" to partition the angles into different categories*

At this point, we may choose between two strategies: we could perform clustering by initially ignoring the amplitude ratios and later merging classes located within the same angular domain if, after merging, a refractory period is indeed visible in the distribution of the intervals between action potentials; alternatively, we can perform clustering separately on the angular domains, merging classes so long as there remains a visible refractory period. For the data of the given example, if we restrict attention to the largest events, a classification based solely on the angular domains will be sufficient, as shown in Figure 8.13.

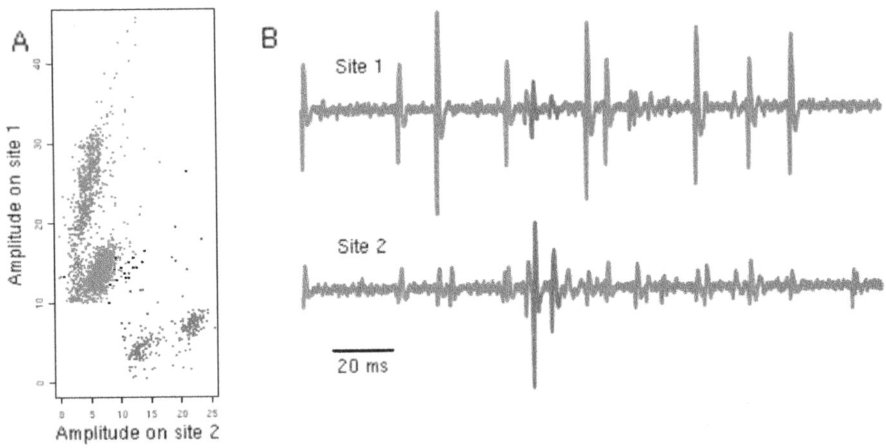

Figure 8.13. *Classification based on amplitude ratios. A) Left section of Figure 8.12, with points colored by amplitude ratio. Note that only spikes with peak amplitude larger than 10 at one or more sites have been retained. B) Identical to Figure 8.11 with spikes colored by amplitude ratio. The cell firing in pairs has been correctly identified (in blue), and so has the cell firing in bursts (in brown) and the cell firing "constantly" (in orange). For a color version of this figure, see www.iste.co.uk/clerc/interfaces1.zip*

8.3.10. *Sampling jitter (1984)*

One specific difficulty arises with the technique of data sampling. The data are physically saved in the form of sequences (or vectors) of amplitudes – values of the amplitude at uniformly separated points in time – whereas the original true data were *continuous*. In an ideal situation, without any recording noise, we would expect the position of two consecutive action potentials generated by the same neuron to be shifted relative to the sampling times (technically, we would usually use the term *phase* rather than position for this), as shown in Figure 8.14(a). If analysis is performed directly on the sampled data, for example an attempt to resolve superposition by subtracting the closest-fitting template (Figure 8.8), new events may be unintentionally introduced as a result, as shown in Figure 8.14(b1). In this example, the event sampled at the bottom of Figure 8.14(a) was used as a template and subtracted from the event sampled at the top of Figure 8.14(a). The peak amplitude of the difference is equal to five times the standard deviation of the noise, which means that it would be identified as a new spike, as our detection threshold was chosen to be four times the standard deviation of the noise for these data. Another way of visualizing the consequences of sampling jitter is by simulating, using a continuous template – more precisely a template defined by a continuous *function* – noisy sampled data with and without jitter. The jitter is simulated by a uniformly distributed random variable taking values in an interval of $-1/2$ to $+1/2$ of the sampling period. The template is then subtracted from the simulated data, and the sum of the squares of the residues is calculated (as we did earlier for Figure 8.5(b)). The distribution of the sum of the squares of the residues is shown in Figure 8.14(b2). We see that jitter can have an effect on the variability that is of the same order of the effect of noise. This effect depends on the sampling rate and on the shape of the template, as explained in [POU 14]. The usual strategy for counteracting this problem is to sample at high frequencies, but this is only feasible when relatively few channels are recorded simultaneously. Another option is to numerically resample by application of the Nyquist–Shannon theorem [POU 02]. It is also possible to effectively correct for jitter using the method suggested in 1984 by McGill and Dorfman [MCG 84] – using Fourier transforms – or using a Taylor–McLaurin series expansion [POU 14]. Finally, for purposes of resolving superposition – in the author's experience – it seems to be the case that the effect of jitter is less noticeable when using filters (section 8.3.8) as compared to subtraction-based methods (section 8.3.6), although these apparent differences are yet to be documented in a "serious" study.

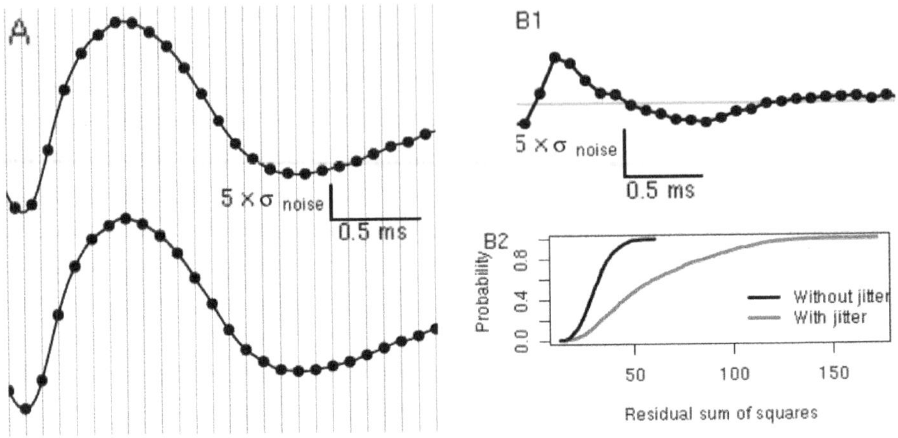

Figure 8.14. *Sampling jitter. A) Two action potentials from the same neuron sampled with two different phases; the points correspond to the digitally sampled amplitudes, and the continuous line corresponds to the data pre-sampling. B1) Illustration of the jitter effect* – without recording noise – *with template matching; here, the template is the sampled version of the figure at the bottom of A, and the event is the sampled version of the figure shown above. The "bottom" amplitudes are subtracted from the top amplitudes (which precede them by half of an amplitude period). B2) A simulation comparing 1,000 events with and without uniform jitter on +/- half of a sampling period with white recording noise following a normal distribution (here, the events were defined by sequences of 30 amplitudes). For a color version of this figure, see www.iste.co.uk/clerc/interfaces1.zip*

8.3.11. *Graphical tools*

Since the late 1980s, there have been spectacular improvements in the computational power of computers, with the introduction of *interactive* methods of visualization, the first and foremost of which is most certainly the program XCLUST developed by Wilson[24]. These methods involve the systematic application of the techniques of dimension reduction as discussed in section 8.3.4; they allow multiple projections to be visualized simultaneously. Thus, instead of working with only the first two principal components (Figure 8.7), we are able to work with four or more, and compare the graphs of the projections onto planes defined by pairs of any two of the

24 The latest versions of this program are available on github: github.com/wilsonlab/mwsoft64.

principal components (first and second, first and third, etc.). Figure 8.15 shows a screenshot of the software package `GGobi` – free of cost and open source[25] – showing an example of one such matrix of projections[26]. This figure attempts to show the interactivity of the program as much as possible. The "active" panel or graph is at the top-left of the figure (with a black border). The little light-blue square is the "paintbrush", which the user is free to move using the mouse. Each point, initially magenta colored, becomes blue on the active graph once it is selected by the blue square as well as the corresponding points on all of the other graphs in the matrix. The technique of coloring in parallel equivalent points on multiple graphs is called *brushing*; see [CLE 93, p. 294] and [COO 07]. Today, this is how most spike sorting is performed. The program that we are showcasing here, `GGobi`, is capable of providing even more sophisticated (and extremely useful) dynamic visualizations, such as the "grand tours" introduced by Asimov [ASI 85]. In our experience, although `GGobi` is not sufficient[27] for spike sorting, it is, nevertheless, the most important software package for this task.

8.3.12. *Automatic clustering*

With the development of interactive graphical methods since the late 1980s, the development, or the adoption, of automatic or semiautomatic clustering methods has been the greatest focus of spike sorting "methodologists". Indeed, the problem with the methods presented up to this point is that they require "significant" effort from the researcher performing data analysis. "Template matching" (section 8.3.3) and "filtering" (section 8.3.8) require the templates and filters to be estimated, and the methods combining dimension reduction and clustering (sections 8.3.4, 8.3.5 and 8.3.11) require the classes or event groups to be defined directly by the user. These "heavy" tasks have the following two effects on the analysis:

25 Available for download free of cost for `Linux`, `Windows` and `Mac` at http://www.ggobi.org/.

26 This figure was prepared using the locust dataset, but unlike Figures 8.6 and 8.7, all of the events and all four recording sites were included – for clarity, the previous figures were prepared with subsets of the events from one single recording site.

27 Because resolving superposition is not possible, at least not easily, after performing dimension reduction.

1) the analysis becomes time intensive;

2) the analysis becomes difficult to reproduce[28].

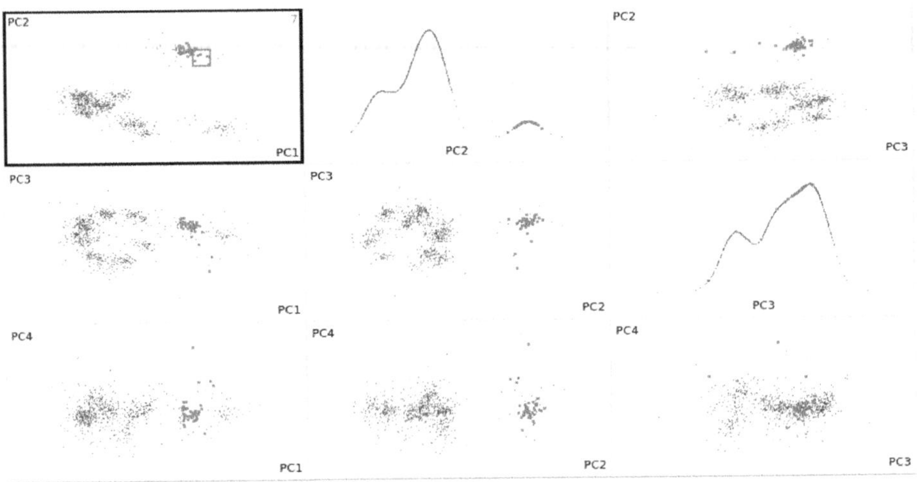

Figure 8.15. *Matrix of scatter plots showing the projections of a set of events onto the planes defined by pairs taken from the set of the first four principal components: a "screenshot" of the software package `GGobi`. The "diagonal" graphs (first row, second column and second row, third column) are the smooth estimates of the densities of the projections of the events onto the second (first row, second column) and third (second row, third column) principal components. For a color version of this figure, see www.iste.co.uk/clerc/interfaces1.zip*

There is therefore a strong demand for automatic methods, which has in the past inspired a large number of publications (and still does to this day). At the risk of angering a fair few of our colleagues, we wish to venture the opinion that most of the obstacles encountered in the context of (the clustering stage of) spike sorting are addressed in sufficient depth by the two most common statistical methods for this type of problem:

28 Reproducibility fails at two different levels: two different people analyzing the same dataset will usually not define the same classification as illustrated in [HAR 00]; and one *same* person analyzing the *same* dataset 6 months later will usually not define the same classification twice.

1) the *k-means* algorithm[29];

2) *Gaussian mixture models* (or GMM)[30] modified by the *expectation–maximization* or *[EM] algorithm*)[31].

The k-means algorithm is easy to specify (and implement):

– *Choice of number of components*: the number of classes k to include in the model[32] is chosen by the user;

– *Initialization*: k events, which we shall call *centroids*, are chosen at random among the n observed events;

– *Distance calculation*: the (Euclidean) distance of each event from each of the k centroids is calculated;

– *Event assignment*: each event is assigned to the centroid to which it is closest;

– *Centroid update*: the *updated* position of each centroid is calculated as the average of the events that "belong" to that centroid;

– *Iteration*: return to the *distance calculation* step until a maximal number of iterations – chosen beforehand – is reached, or another stopping condition[33] is satisfied.

– *Results*: the final values of the centroids are the "templates", the final assignments yield the classification and the "total variance"[34] is calculated.

This procedure is repeated multiple times (10–50 times) with *different initializations*; the final result is taken to be the instance with the smallest final total variance. The way that the algorithm works is illustrated in Figure 8.16. The data are from Figure 8.15, but to make the illustration easier to read, one

29 See: https://en.wikipedia.org/wiki/K-means_clustering.

30 See: https://en.wikipedia.org/wiki/Mixture_model.

31 See: https://en.wikipedia.org/wiki/Expectation-maximization_ algorithm.

32 In practice, observing the data using "dynamic" modes (rotations and "grand tours") in GGobi allows k to be chosen. We will discuss automatic methods at a later point.

33 An example of a stopping condition is when all distances between two between consecutive values for each of the centroids are below a chosen threshold.

34 Each centroid is subtracted from each of its assigned events (vector subtraction) and the squares of the (Euclidean) norms of these differences are summed. This sum is denoted the "total variance".

single projection (onto the plane defined by the first and third principal components) was used and the events from three of the 10 neurons – which were identified when the analysis was performed "properly" – were omitted.

The EM algorithm for a GMM adds an extra layer of formalism to the k-means algorithm: a probabilistic model of data generation is therefore *explicitly* assumed. Each observation $\mathbf{y}_1, \mathbf{y}_2, \dots, \mathbf{y}_n$ is viewed as the realization of a random variable $\mathbf{Y} \in \mathbb{R}^p$ whose distribution is known up to a *finite* number of parameters. In the case of a *Gaussian mixture*, the density of $\mathbf{Y}$ may be written as:

$$p\left(\mathbf{Y} = \mathbf{y}; \theta_k\right) = \sum_{j=1}^{k} \pi_j \, \phi(\mathbf{y}; \mu_j, \Sigma_j), \qquad [8.8]$$

where θ_k is the set of model parameters,

$$\theta_k = \{\pi_j, \mu_j, \Sigma_j\}_{j=1,\dots,k}, \; 0 \leq \pi_j \leq 1, \; \sum_{j=1}^{k} \pi_j = 1, \qquad [8.9]$$

and where $\phi(\;;\mu, \Sigma)$ is the density of a multidimensional normal (or Gaussian) distribution:

$$\phi(\mathbf{y}; \mu, \Sigma) = \frac{1}{(2\pi)^{p/2}|\Sigma|^{1/2}} \exp\left(-\frac{1}{2}(\mathbf{y} - \mu)^T \Sigma^{-1} (\mathbf{y} - \mu)\right), \qquad [8.10]$$

where μ is the mean, a vector in $\mathbb{R}^p$, Σ is the covariance matrix[20], $|\Sigma|$ is the determinant of Σ and the superscript T is the *transpose*. The unknowns of the mixture distribution are the weights π_j – there are only $k - 1$ independent values – the k means μ_j and the k covariance matrices Σ_j. With this setup, the EM algorithm for a GMM is only slightly more complicated than the k-means algorithm. In the general case where each neuron/aggregation has its own covariance matrix, it may be stated as follows:

– *Initialization*: k events are randomly chosen from the n observed events, which are taken as the k $\mu_j^{(0)}$. The $\pi_j^{(0)}$ are typically all initialized with identical values equal to $1/k$, and the $\Sigma_j^{(0)}$ are also initialized identically as diagonal matrices with elements equal to the variance of the noise;

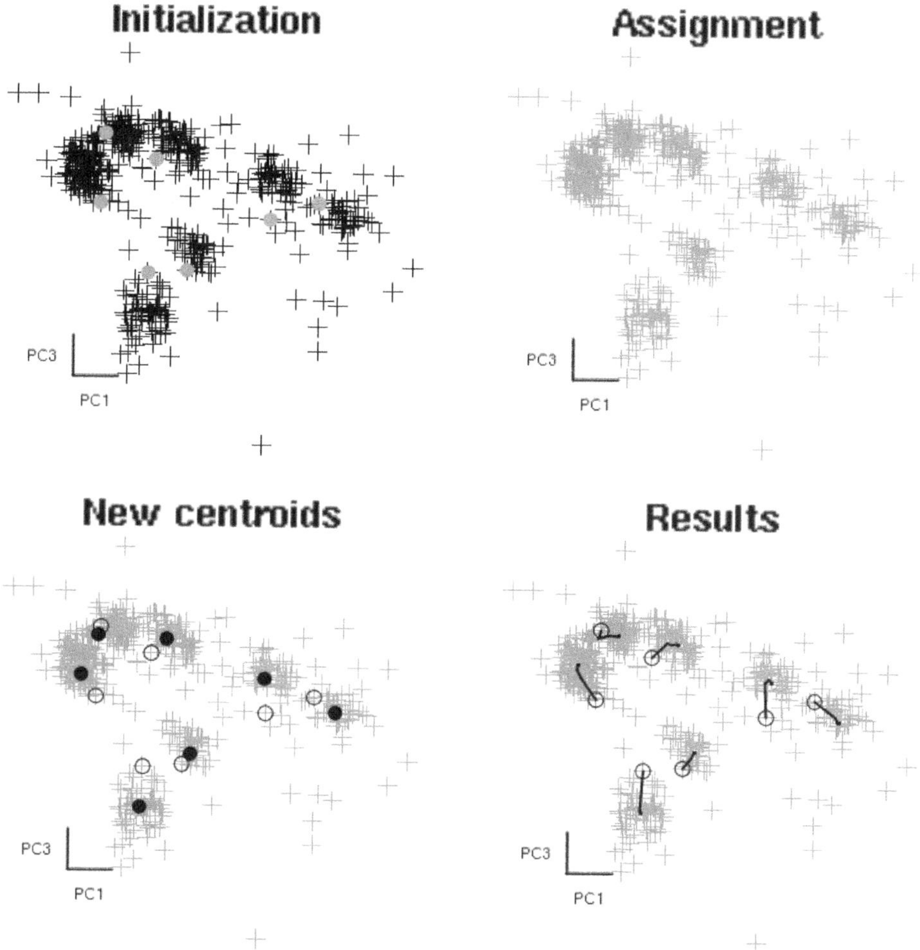

Figure 8.16. *K-means algorithm. Initialization: seven events (colored discs) were randomly chosen from the set of events (black crosses). Assignment: the distance between each centroid and each event was calculated, and each event was assigned to the nearest centroid. Note how the concentration of points centered around the sky blue section is partitioned into segments such that its edges are assigned to its neighbors, colored magenta and turquoise. New centroids: the updated positions of the centroids (black discs) are calculated; the old positions are shown as circles. Results: after 20 iterations of the algorithm, the final assignments are obtained. The trajectories of the centroids are shown in black, and the initial positions are shown as circles. For a color version of this figure, see www.iste.co.uk/clerc/interfaces1.zip*

– *Calculation of relative likelihoods*: the relative likelihood $p_{i,j}$ that the jth of the k centroids generated the ith event is calculated:

$$p_{i,j} = \pi_j^{(l)} \, \phi(\mathbf{y}_i; \mu_j^{(l)}, \Sigma_j^{(l)})$$

– *Assignment of responsibilities*: the "responsibility" $t_{i,j}$ of each of the centroids j for each of the events i is obtained by normalizing the relative likelihoods:

$$t_{i,j} = p_{i,j} / \sum_{m=1}^{k} p_{i,m}$$

– *Update of parameters*: new parameter values are obtained for each centroid by averaging each event weighted by the responsibility of the corresponding centroid:

$$\pi_j^{(l+1)} = \sum_{i=1}^{n} t_{i,j}/n \,,$$

$$\mu_j^{(l+1)} = \left(\sum_{i=1}^{n} t_{i,j} \, y_i \right) / \sum_{i=1}^{n} t_{i,j}$$

and

$$\Sigma_j^{(l+1)} = \left(\sum_{i=1}^{n} t_{i,j} \, (y_i - \mu_j^{(l+1)})(y_i - \mu_j^{(l+1)})^T \right) / \sum_{i=1}^{n} t_{i,j}$$

– *Iteration*: return to step *calculation of relative likelihoods* until a maximal number of iterations – chosen beforehand – has been performed, or another stopping condition is satisfied;

– *Results*: the results are given by the latest values for the parameters and the responsibilities.

Analogously to the k-means algorithm, this procedure is repeated multiple times (from 10–50 times) with *different initializations*; the final result chosen from the instance of the algorithm that produces the greatest final probability density – or likelihood, as explained in the next section – for the given dataset[35]. The μ_j give estimates for the centroids. The responsibilities may be utilized in two different ways: the first approach is to assign each event to the centroid with the greatest responsibility for that event; the second is to record the responsibilities as they are and perform all subsequent estimations (histograms of intervals between action potentials, cross-correlograms between neurons, etc.) by taking averages weighted by the responsibilities, as explained in section 5.4 of [POU 05]. Possible (and widely used) simplifications of the GMM as specified above include taking the π_j to be identical for all neurons – which amounts to assuming that they all fire with identical frequency – and taking the Σ_j to be identical – which amounts to assuming that each individual neuron always generates spikes of the same shape or template; in other words, there are no dynamic shape profiles (sections 8.3.7 and 8.3.9). By combining these last two constraints, we obtain a version of the *k*-means algorithm that allows for partial assignments in the assignment step. On the dataset in Figure 8.16, regardless of the version of GMM chosen, the algorithm produces a classification identical to the classification given by *k*-means, assuming that the same number of classes/neurons is used. In practice, GMM with EM is preferred over *k*-means when the concentrations of points visualized with GGobi have different shapes and, most importantly, when they partially overlap. When they overlap, estimating the position of the centroids (the μ of a MMG) will be more reliable, which is important when using these methods of automatic clustering as a preamble to a classification based on template matching (section 8.3.3) or filtering (section 8.3.8).

The theoretical basis for methods of automatically choosing the number of classes – the k parameter in the above – is the concept of penalized likelihood, and are discussed in Chapter 7 of [HAS 09]. The *likelihood* is simply the probability density of the observations (and the log-likelihood is its logarithm), except that the role of the observations and the parameters have been switched; for example, in the case of the GMM considered above:

$$l(\theta_k; \mathbf{y}_1, \ldots, \mathbf{y}_n) = \sum_{i=1}^{n} \log \left\{ \sum_{j=1}^{k} \pi_j \, \phi(\mathbf{y}_i; \mu_j, \Sigma_j) \right\} . \qquad [8.11]$$

35 We would usually calculate the final log-likelihood instead: $\sum_{i=1}^{n} \log \left(\sum_{j=1}^{k} \pi_j \, \phi(\mathbf{y}_i; \mu_j, \Sigma_j) \right)$.

The log-likelihood is *a function of the parameters, assuming that the data are fixed.* A major mathematical result in this area of statistics is that by choosing the estimator $\widehat{\theta}$ of θ to be the argument that maximizes equation [8.11] *for a fixed number of classes*, we achieve an "optimal" value[36]. The fact that the sequence $\theta^{(l)}$ generated by the EM algorithm converges to $\widehat{\theta}$ (assuming some fairly general conditions) is another important mathematical result. Now, in the case of a GMM, we can immediately see that as k increases and the diagonal elements of the Σ_j decrease, the log-likelihood becomes infinite, for example if we take the number of classes to be equal to the number of observations and set $\pi_j = 1/n$ and the μ_j equal to the observations (one centroid per observation). In other words, if we attempt to maximize the likelihood while allowing the number of classes to vary, without setting a lower bound (> 0) for the diagonal elements of the Σ_j, then the likelihood is maximized by a model with as many classes as there are observations, where each centroid is equal to one of the observations and where the covariance matrices are degenerate with zeroes along the diagonal. If the data are being continually recorded, it is in principle possible to estimate the covariance matrix of the noise (similarly to [POU 02]) and to use this estimation as a constraint for the covariance matrices in each of the classes: the elements of the covariance matrices must be greater than or equal to the corresponding elements in the covariance matrix of the noise. Interestingly, this approach does not seem to have ever been pursued. Instead, more general statistical methods are typically used; these methods do not assume that it is possible to independently estimate the noise level, penalizing the likelihood by a term proportional to the "complexity" of the model – in other words, the number of parameters. This approach leads us to minimize the *Akaike information criterion* (AIC)[37]:

$$\mathrm{AIC}(k) = -2\,l(\widehat{\theta}_k) + 2\,d\,, \qquad [8.12]$$

36 Optimal in the sense that if the data were indeed generated by a mixture of Gaussian models, and if the number of observations n tends to infinity, then the random variable $\widehat{\theta}$ will converge to θ and has the smallest possible variance.

37 See: https://en.wikipedia.org/wiki/Akaike_information_criterion.

where $\widehat{\theta}_k \in \mathbb{R}^d$ maximizes equation [8.11] and d is the dimension of the parameter space, which is a function of k. Another even more commonly used criterion is the *Bayesian information criterion* (BIC)[38]:

$$\mathrm{BIC}(k) = -2\,l(\widehat{\theta}_k) + d\,\log n\,. \quad [8.13]$$

In the light of the discussion above, as k increases, so too does $l(\widehat{\theta}_k)$, and the first terms of the AIC and the BIC decrease; it is clear that the terms $2d$ and $d\log n$ will counteract this decrease, as they themselves increase with k. Thus, the BIC penalizes complex models more strongly than the AIC. In practice, both criteria overestimate the number of classes/neurons. This is largely due to the fact that events can overlap when clustering is performed (sections 8.3.6 and 8.3.8); these instances of superposition are not correctly accounted for in mixed models (see the remark about this at the end of section 8.3.11). It would clearly be desirable to perform a comparison of these models based on complete datasets specifically including information about instances of superposition, not just at the clustering stage, but this has not yet been pursued to our knowledge.

8.4. Recommendations

We would like to conclude this chapter with various recommendations, ranging from general tips to more specific advice. First of all, a piece of advice that holds in much more generality than simply the field of spike sorting: readers should *never* use methods that they do not understand. In the context of spike sorting, and for data analysis in neurophysiology in general, *an excellent way to understand a method is to program it.* Today, there are many generalistic environments – or "ecosystems" as they are increasingly called by programmers – for data analysis: Python[39], R[40]; these ecosystems provide a platform for the methods discussed in the literature to be rapidly and easily implemented. As an example, we invite the reader to refer to

38 See: http://en.wikipedia.org/wiki/Bayesian_information_criterion.

39 Official website: https://www.python.org/, with the additional packages `Numpy`, `Scilab`, `Matplotlib` (the Web site http://www.scipy.org/ can serve as an entry point).

40 Official website: http://www.r-project.org/.

the analysis of the two datasets used in this chapter in R and Python[41]. This advice should not be understood to imply that for "serious" analysis authors should necessarily reprogram all methods for themselves; clearly, for algorithms such as k-means or EM for GMM, effective and *well-tested* code is available and should be preferred. Nevertheless, a good understanding of these two algorithms may be easily obtained even just by programming simple versions of them. The advantage of our recommended approach, which will prove massive in the medium or long term, is that it enables data analysts to unshackle themselves from the methods provided by manufacturers (generally amplifier manufacturers); in our experience, these methods are opaque and insufficiently adaptable[42]. After more than 15 years of working in spike sorting (among other things, thankfully), on various different species (rats, mice, monkeys, locusts, beetles, bees), various different tissue types (cerebellum, hippocampus, neocortex, antennal lobe, etc.) and with various types of electrode, we have learned that certain key stages: filtering, event detection, clustering methods must be adapted to suit the tissue type[43]. Once these adjustments have been made, it is very straightforward, in environments with the right support such as Python or R, to write a script with very few parameters (or even no parameters) that can perform the entire sorting process for a given dataset. With this is mind, we *strongly* recommend using a modernized version of the approaches used until the mid-1980s:

1) The use the first minute of recording (or the first few minutes) to estimate the templates (section 8.3.3) with GGobi followed by k-means or EM for a GMM if k-means does not produce satisfactory results, or *bagged clustering* [LEI 99] if both of these options are not satisfactory;

2) Establish a classification by template matching (section 8.3.3) or using filters (sections 8.3.8) *after resolving instances of superposition* (section 8.3.6), accounting for sampling jitter (section 8.3.10) and dynamic amplitude

41 See the page dedicated to spike sorting on the author's website: http://xtof.perso.math.cnrs.fr/sorting.html, examples of analysis may be found at the bottom of the page.

42 These two problems prompted the author to first begin programming his own methods.

43 This holds for species/tissue pairs; thus different methods are used for recordings in the antennal lobe (the insect equivalent of the olfactory bulb for vertebrates) for locusts and for beetles.

profiles[44] (sections 8.3.7 and 8.3.9). This classification should be based solely on a recording period that is "short" compared to the electrode drifting timescale, which triggers changes in the templates and should not be performed over the whole of the recording;

3) Correct the templates for drifting, if necessary, and establish a classification for the next period of recording.

In general, the use the median instead of the mean – this is particularly important for template estimation – and the median of the absolute value of the deviations with respect to the median instead of the standard deviation (in short, the *median absolute deviation*); these two estimators are "robust"[45]. These two recommendations are a lot more important than they might seem; in practice, they produce a considerable improvement in the reliability of the results, and not just for spike sorting. Finally, the literature on spike sorting, and on the analysis of neurophysiological data in general, is relatively opaque; the author firmly believes that this problem could be reduced, or perhaps completely solved, if users/developers gave *unrestricted access to their data and their programs*, or in other words if they conducted their research so as to be reproducible [STO 14, DEL 12]: this is the best way to achieve both individual and collective progress.

8.5. Bibliography

[ADR 22] ADRIAN E.D., FORBES A., "The all-or-nothing response of sensory nerve fibres", *Journal of Physiology*, vol. 56, no. 5, pp. 301–330, 1922.

[ANT 00] ANTIC S., WUSKELL J.P., LOEW L. *et al.*, "Functional profile of the giant metacerebral neuron of Helix aspersa: temporal and spatial dynamics of electrical activity in situ", *The Journal of Physiology*, vol. 527, no. 1, pp. 55–69, 2000.

[ASI 85] ASIMOV D., "The grand tour: a tool for viewing multidimensional data", *SIAM Journal on Scientific and Statistical Comput*ing, vol. 6, no. 1, pp. 128–143, 1985.

[BED 04] BEDARD C., KROGER H., DESTEXHE A., "Modeling extracellular field potentials and the frequency-filtering properties of extracellular space", *Biophysics Journal*, vol. 86, no. 3, pp. 1829–1842, 2004.

[BRE 09] BRÉMAUD P., *Initiation aux Probabilités et aux chaînes de Markov*, Springer, Berlin Heidelberg, 2009.

44 We still need an effective algorithm for resolving superposition in the presence of dynamic amplitude profiles.

45 See: http://en.wikipedia.org/wiki/Robust_statistics.

[BUZ 04] BUZSÁKI G., "Large-scale recording of neuronal ensembles", *Nature Neurosciences*, vol. 7, no. 5, pp. 446–451, 2004.

[CAL 73] CALVIN W.H., "Some simple spike separation techniques for simultaneously recorded neurons", *Electroencephalography and Clinical Neurophysiology*, vol. 34, no. 1, pp. 94–96, 1973.

[CAN 10] CANEPARI M., ZECEVIC D., *Membrane Potential Imaging in the Nervous System: Methods and Applications*, Springer, 2010.

[CHA 99] CHAPIN J., MOXON K., MARKOWITZ R. *et al.*, "Real-time control of a robot arm using simultaneously recorded neurons in the motor cortex", *Nature Neuroscience*, vol. 2, no. 7, pp. 664–670, 1999.

[CHA 04] CHAPIN J., "Using multi-neuron population recordings for neural prosthetics", *Nature Neuroscie*nce, vol. 7, no. 5, pp. 452–455, 2004.

[CLE 93] CLEVELAND W.S., *Visualizing Data*, Hobart Press, NJ, 1993.

[COO 07] COOK D., SWAYNE D.F., *Interactive and Dynamic Graphics for Data Analysis. With R and GGobi*, Springer, Springer Science+Business Media, LLC, New York, 2007.

[DEL 06] DELESCLUSE M., POUZAT C., "Efficient spike-sorting of multi-state neurons using inter-spike intervals information", *Journal of Neuroscience Methods*, vol. 150, no. 1, pp. 16–29, 2006.

[DEL 12] DELESCLUSE M., FRANCONVILLE R., JOUCLA S. *et al.*, "Making neurophysiological data analysis reproducible. Why and how?", *Journal of Physiology (Paris)*, vol. 106, nos. 3–4, pp. 159–170, 2012.

[DON 08] DONG Y., MIHALAS S., QIU F. *et al.*, "Synchrony and the binding problem in macaque visual cortex", *Journal of Vision*, vol. 8, no. 7, pp. 1–16, 2008.

[EYZ 55] EYZAGUIRRE C., KUFFLER S.W., "Further study of soma, dendrite, and axon excitation in single neurons", *Journal of General Physiol*ogy, vol. 39, no. 1, pp. 121–153, 1955.

[FAT 57] FATT P., "Electric potentials occurring around a neurone during its antidromic activation", *Journal of Neurophysiol*ogy, vol. 20, no. 1, pp. 27–60, 1957.

[GEO 86] GEORGOPOULOS A., SCHWARTZ A., KETTNER R., "Neuronal population coding of movement direction", *Science*, vol. 233, no. 4771, pp. 1416–1419, 1986.

[GER 64] GERSTEIN G.L., CLARK W.A., "Simultaneous studies of firing patterns in several neurons", *Science*, vol. 143, no. 3612, pp. 1325–1327, 1964.

[GLA 68] GLASER E., MARKS W., "On-line separation of interleaved neuronal pulse sequences", in ENSLEIN K., (ed.), *Data Acquisition and Processing in Biology and Medicine*, Pergamon, 1968.

[GLA 76] GLASER E.M., RUCHKIN D.S., *Principles of Neurobiological Signal Analysis*, Academic Press, New York, 1976.

[GOL 74] GOLDSTEIN S.S., RALL W., "Changes of action potential shape and velocity for changing core conductor geometry", *Biophysics Journal*, vol. 14, no. 10, pp. 731–757, 1974.

[GRA 95] GRAY C.M., MALDONADO P.E., WILSON M. *et al.*, "Tetrodes markedly improve the reliability and yield of multiple single-unit isolation from multi-unit recordings in cat striate cortex", *Journal of Neuroscience Methods*, vol. 63, nos. 1–2, pp. 43–54, 1995.

[GRO 70] GROVER F.S., BUCHWALD J.S., "Correlation of cell size with amplitude of background fast activity in specific brain nuclei", *Journal of Neurophysiology*, vol. 33, no. 1, pp. 160–171, 1970.

[HAR 32] HARTLINE H.K., GRAHAM C.H., "Nerve impulses from single receptors in the eye", *Journal of Cellular and Comparative Physiology*, vol. 1, no. 2, pp. 277–295, 1932.

[HAR 00] HARRIS K.D., HENZE D.A., CSICSVARI J. *et al.*, "Accuracy of tetrode spike separation as determined by simultaneous intracellular and extracellular measurements.", *Journal of Neurophysiology*, vol. 84, no. 1, pp. 401–414, 2000.

[HAS 09] HASTIE T., TIBSHIRANI R., FRIEDMAN J., *The Elements of Statistical Learning: Data Mining, Inference, and Prediction*, 2nd ed., Springer, 2009.

[HOD 52] HODGKIN A.L., HUXLEY A.F., "A quantitative description of membrane current and its application to conduction and excitation in nerve", *The Journal of Physiology*, vol. 117, no. 4, pp. 500–544, August 1952.

[HOM 09] HOMMA R., BAKER B.J., JIN L. *et al.*, "Wide-field and two-photon imaging of brain activity with voltage- and calcium-sensitive dyes", *Philosophical Transactions of the Royal Society B: Biological Sciences*, vol. 364, no. 1529, pp. 2453–2467, 2009.

[LEI 99] LEISCH F., Bagged clustering, Working Paper no. 51, Vienna University of Economics and Business, 1999.

[LIN 14] LINDÉN H., HAGEN E., LESKI S. *et al.*, "LFPy: A tool for biophysical simulation of extracellular potentials generated by detailed model neurons", *Frontiers in Neuroinformatics*, vol. 7, no. 41, 2014.

[MAL 81] VON DER MALSBURG C., The correlation theory of brain function, Internal Report, MPI Biophysical Chemistry, available at http://cogprints.org/1380/01/vdM_correlation.pdf, nos. 81–82 1981.

[MCG 84] MCGILL K.C., DORFMAN L.J., "High-resolution alignment of sampled waveforms", *IEEE Transactions on Biomedical Engineering*, vol. 31, no. 6, pp. 462–468, 1984.

[MCN 83] MCNAUGHTON B.L., O'KEEFE J., BARNES C.A., "The stereotrode: A new technique for simultaneous isolation of several single units in the central nervous system from multiple unit records", *Journal of Neuroscience Methods*, vol. 8, no. 4, pp. 391–397, 1983.

[PLO 07] PLONSEY R., BARR R., *Bioelectricity: A Quantitative Approach*, Springer, Springer Science+Business Media, LLC, New York, 2007.

[POG 63] POGGIO G.F., MOUNTCASTLE V.B., "The functional properties of ventrobasal thalamic neurons studied in unanesthetized monkeys", *Journal of Neurophysiology*, vol. 26, pp. 775–806, 1963.

[POU 02] POUZAT C., MAZOR O., LAURENT G., "Using noise signature to optimize spike-sorting and to assess neuronal classification quality", *Journal of Neuroscience Methods*, vol. 122, no. 1, pp. 43–57, 2002.

[POU 05] POUZAT C., "Course 15 – Technique(s) for spike-sorting", *Les Houches*, vol. 80, pp. 729–785, 2005.

[POU 14] POUZAT C., DETORAKIS G.I., "SPySort: neuronal spike sorting with python", *Proceedings of the 7th European Conference on Python in Science (EuroSciPy'14)*, pp. 27–34, 2014.

[PRO 72] PROCHAZKA V., CONRAD B., SINDERMANN F., "A neuroelectric signal recognition system", *Electroencephalography and Clinical Neurophysiology*, vol. 32, no. 1, pp. 95–97, 1972.

[RAL 77] RALL W., "Core conductor theory and cable properties of neurons", *Handbook of Physiology, Cellular Biology of Neurons*, American Physiological Society, Bethesda, MD, 1977.

[ROB 75] ROBERTS W.M., HARTLINE D.K., "Separation of multi-unit nerve impulse trains by a multi-channel linear filter algorithm", *Brain Research*, vol. 94, no. 1, pp. 141–149, 1975.

[ROB 79] ROBERTS W.M., "Optimal recognition of neuronal waveforms", *Biological Cybernetics*, Springer, Berlin/Heidelberg, vol. 35, pp. 73–80, 1979.

[SIM 65] SIMON W., "The real-time sorting of neuro-electric action potentials in multiple unit studies", *Electroencephalography and Clinical Neurophysiology*, vol. 18, no. 2, pp. 192–195, 1965.

[STO 14] STODDEN V., LEISCH F., PENG R.D., *Implementing Reproducible Research*, Chapman & Hall/CRC The R Series/Taylor & Francis, Boca Raton, 2014.

[WES 00] WESSBERG J., STAMBAUGH C.R., KRALIK J.D. *et al.*, "Real-time prediction of hand trajectory by ensembles of cortical neurons in primates", *Nature*, vol. 408, no. 6810, pp. 361–365, 2000.

[WIL 99] WILLIAMS S.R., STUART G.J., "Mechanisms and consequences of action potential burst firing in rat neocortical pyramidal neurons", *The Journal of Physiology*, vol. 521, no. 2, pp. 467–482, December 1999.

[ZEC 89] ZECEVI D., WU J.Y., COHEN L.B. *et al.*, "Hundreds of neurons in the Aplysia abdominal ganglion are active during the gill-withdrawal reflex", *Journal of Neuroscience*, vol. 9, no. 10, pp. 3681–3689, 1989.

9

Statistical Learning for BCIs

This chapter introduces statistical learning and its applications to brain–computer interfaces. We begin by presenting the general principles of supervised learning and discussing the difficulties raised by its implementation, with a particular focus on aspects related to selecting sensors and multisubject learning. This chapter also describes in detail how a learning approach may be validated, including various metrics of performance and optimization of the hyperparameters of the considered algorithms.

We invite the reader to experiment with the algorithms described here: the illustrative experiments included in this chapter may be reproduced using a Matlab/Octave toolbox[1], which contains the implementation details of the various different methods.

9.1. Supervised statistical learning

The goal of supervised learning is to construct a predictor function that assigns a label to any given example; this predictor function is constructed from labeled examples that provide a basis for this training process. The predictor function is obtained by optimizing a certain criterion, which includes a term for the empirical risk (the behavior of the function on the given examples) and a regularization term (which guarantees the behavior of the function on new examples).

Chapter written by Rémi FLAMARY, Alain RAKOTOMAMONJY and Michèle SEBAG.

1 Matlab/Octave toolbox: https://github.com/rflamary/mltool.

This section will describe the principles of supervised statistical learning, as well as the two most important algorithms in the current state of the art. The reader should refer to Lotte *et al.* [LOT 07] for a more detailed account of the state of the art.

9.1.1. *Training data and the predictor function*

The objective of supervised statistical learning is to estimate a predictor function $f(\mathbf{x}) : \mathbb{R}^d \rightarrow \mathcal{Y}$ which, given an observation $\mathbf{x} \in \mathbb{R}^d$, predicts a label $y \in \mathcal{Y}$ [HAS 01, DUD 99]. The coordinates of the vector $\mathbf{x}$ are the descriptive features extracted from the observation $\mathbf{x}$ (see Chapters 7, 8 and 10). Traditionally, we distinguish between methods of classification (or discrimination), for which the label y is nominal (for example $\mathcal{Y} = \{-1, 1\}$ is a problem of binary classification) and methods of regression, for which the label y is real ($\mathcal{Y} = \mathbb{R}$).

In practice, the function $f(\cdot)$ is estimated from a set of n training examples $\mathcal{E} = \{(\mathbf{x}_i, y_i), \mathbf{x}_i \in \mathbb{R}^d, y_i \in \mathcal{Y}, i = 1, \ldots, n\}$. The target function $f(\cdot)$ must perform well on the given data (predict well the labels y_i of the training data $\mathbf{x}_i$) but must also, more importantly, perform well on future examples $\{\mathbf{x}_j, y_j\}$: we say that the function must *generalize* these data.

In this chapter, we will restrict attention to linear predictor functions, defined by:

$$f(\mathbf{x}) = \sum_{j=1}^{d} w_j x_j + b = \mathbf{x}^\top \mathbf{w} + b \qquad [9.1]$$

where $\mathbf{w} \in \mathbb{R}^d$ is a vector of weights, w_j is the weight of the jth coordinate of the decision and function and $b \in \mathbb{R}$ is the bias (constant). Linear predictor functions are the most commonly employed category of functions for brain–computer interfaces due to the fact that they are easy to interpret (the weight w_j may be thought of as the impact of the jth coordinate) and they are straightforward to train [BLA 06, TAN 12]. Note that linear functions (equation [9.1]) return real values; in the case of problems of binary classification, the predicted class is obtained by taking the sign of the function $f(\cdot)$ and not its value.

Depending on the BCI context, the supervised learning problem is either a problem of classification or regression. For example, tasks of motor imagery are generally problems of classification (Figure 9.1, left). The coordinates of $\mathbf{x}$, for example, correspond to the values of the power in a frequency band after CSP-type spatial filtering [LOT 11]. Similarly, a P300 speller task aims to detect (classify) event-related potentials directly in the signal. In this case, the features are, for example, given by the various different filtered and subsampled EEG signals [RAK 08]. The predictor function, or classifier, partitions the space into regions each of which corresponds to one class; the separating hyperplane is defined by the vector $\mathbf{w}$ (normal to the separating hyperplane). In contrast, tasks of movement prediction generally define regression problems; the objective could, for example, be to predict a real value that characterizes the position of a limb from the ECoG measurements (see Figure 9.1, right).

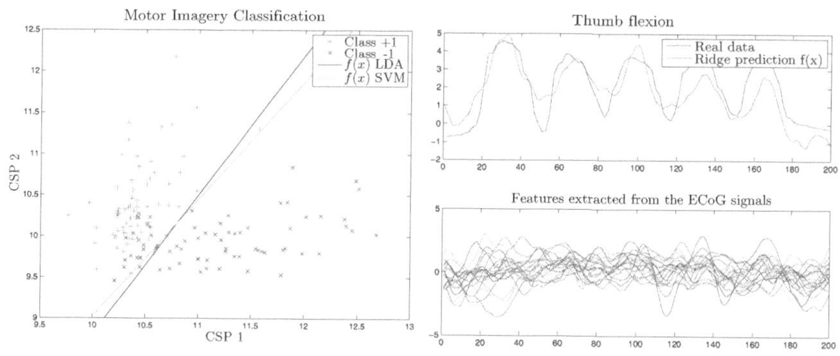

Figure 9.1. *Supervised learning: Data and prediction functions. Left: classification, motor imagery data with two CSP filters and a LDA classifier that partitions the space. Right: Regression, ECoG data with ridge regression. For a color version of this figure, see www.iste.co.uk/clerc/interfaces1.zip*

9.1.2. *Empirical risk and regularization*

As mentioned above, the objective of supervised statistical learning is to make as few mistakes as possible over the full set of possible data (consisting of both the training data and future data), or to be more precise, to minimize the cost of all errors. The *empirical risk* is the average cost of the errors committed on the training data:

$$R_{\text{emp}}(f) = \frac{1}{n} \sum_{i=1}^{n} L(y_i, f(\mathbf{x}_i)) \qquad [9.2]$$

where $L(y, f(\mathbf{x}))$ is the cost of replacing the label y with the prediction $f(\mathbf{x})$ for x. The loss function L also plays an important role in performance metrics (section 9.3).

For classification, the best-known loss function is the $0 - 1$ cost function (Figure 9.2, left, shown in black); the cost of the error is 1 if the predictor function has the same sign as the expected class y, and 0 otherwise. However, the training problem defined by this cost function is difficult to solve: it requires the optimization of a non-differentiable, non-convex function. Other loss functions, such as the hinge loss [VAP 98, CHA 07] or the logistic cost [TOM 07] are therefore often preferred (Figure 9.2, left, and Table 9.1). Another alternative is the sigmoid cost function, traditionally used for neural networks.

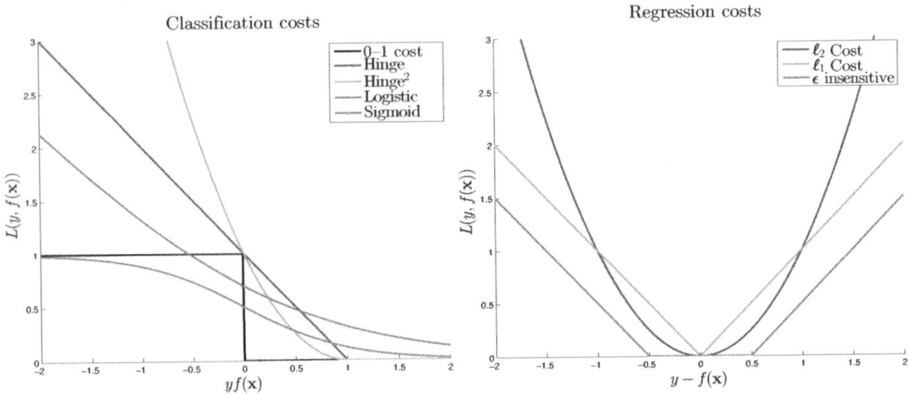

Figure 9.2. *Illustrations of a selection of different loss functions. Left: Classification. Right: Regression. For a color version of this figure, see www.iste.co.uk/clerc/interfaces1.zip*

For regression, the objective is to predict a real value y. The error is traditionally measured by the absolute value of the difference between y and

$f(\mathbf{x})$ [DRA 81] (ℓ_1 loss function) or its square (ℓ_2 loss function). Least squares or regularized least squares (ridge) regression uses the ℓ_2 loss function or mean square error. Another possible loss function is the ϵ-insensitive cost, which is equal to zero whenever the absolute value of the error is less than ϵ [SMO 04]. Using the absolute value (ℓ_1 or ϵ-insensitive) instead of the square (ℓ_2) of the error makes the process more robust against outliers (the impact of a term with a large error is smaller with ℓ_1 than with ℓ_2).

Classification cost	$L(y, f(\mathbf{x}))$
0-1 cost	$\mathbf{1}_{yf(\mathbf{x})<0}$
Hinge	$\max(0, 1 - yf(\mathbf{x}))$
Hinge^2	$\max(0, 1 - yf(\mathbf{x}))^2$
Logistic	$\log(1 + \exp(-yf(\mathbf{x})))$
Sigmoid	$(1 - \tanh(yf(\mathbf{x})))/2$

Reg. cost	$L(y, f(\mathbf{x}))$
ℓ_2 cost	$(y - f(\mathbf{x}))^2$
ℓ_1 cost	$\lvert y - f(\mathbf{x})\rvert$
ϵ-insensitive cost	$\max(0, \lvert y - f(\mathbf{x})\rvert - \epsilon)$

Table 9.1. *Illustrations of loss functions (Figure 9.2); Left: Classification; Right: Regression*

Note that in general minimizing the empirical risk is not sufficient (equation [9.2]) due to the phenomenon of *overfitting*: overfitting is when the function $f(\cdot)$ makes very few errors on the training data but produces high levels of error on subsequent data. This phenomenon occurs frequently when the number of examples n in the training data is small compared to the complexity of the examples (e.g. the number d of attributes). To avoid overfitting, we must limit both the empirical risk and the complexity of the function $f(\cdot)$ by introducing a regularization term $\Omega(f)$ [VAP 98]. The optimization problem therefore becomes:

$$\min_f \quad \frac{1}{n}\sum_{i=1}^{n} L(y_i, f(\mathbf{x}_i)) + \lambda\Omega(f) \qquad [9.3]$$

where $\lambda > 0$, the weight of the regularization term, is a parameter of the algorithm that requires validation (see section 9.4). When $f(\cdot)$ is linear, the ℓ_2 norm of the gradient $\mathbf{w}$ of $f(\mathbf{x})$ provides a good measure $\Omega(f)$ of the complexity, with $\Omega(f) = \sum_{i=1}^{d} w_i^2$. This type of regularization is used in practice for support vector machines (SVM) and ridge regression (section 9.1.3). Another possible measure of complexity is the ℓ_1 norm of $\mathbf{w}$, which has the advantage of implicitly selecting features, or sensors in the context of BCIs, at the cost of introducing a non-differentiable term into the training criterion (section 9.2).

Depending on the nature of the optimization criterion (equation [9.3]), the solution can be found explicitly (e.g. by solving a linear system for ridge regression or linear discriminant analysis [LDA]), or by optimization methods such as the gradient descent method.

9.1.3. *Classical methods of classification*

This section will present two algorithms of linear classification, LDA and SVMs. These algorithms have been successfully applied to brain–computer interfaces, in particular in connection with the software package OpenVibe.

9.1.3.1. *Linear discriminant analysis*

LDA is a Bayesian approach that assumes that the positive (and negative) datapoints follow normal distributions in $\mathbb{R}^d$ given by $\mathcal{N}(\boldsymbol{\mu}_+, \boldsymbol{\Sigma})$ and $\mathcal{N}(\boldsymbol{\mu}_-, \boldsymbol{\Sigma})$), where $\boldsymbol{\mu}_+$ and $\boldsymbol{\mu}_-$ are the means of the normal distributions and $\boldsymbol{\Sigma}$ is the covariance matrix (which is assumed to be identical for both classes).

The decision function $f(\cdot)$ is trained by identifying $\boldsymbol{\mu}_+$, $\boldsymbol{\mu}_-$ and $\boldsymbol{\Sigma}$ and maximizing the likelihood of the training data. $f(\cdot)$ is linear with

$$\mathbf{w} = \boldsymbol{\Sigma}^{-1}(\boldsymbol{\mu}_+ - \boldsymbol{\mu}_-), \qquad b = -\mathbf{w}^\top(\boldsymbol{\mu}_+ + \boldsymbol{\mu}_-)/2 \qquad [9.4]$$

The parameters $\boldsymbol{\Sigma}, \boldsymbol{\mu}_+, \boldsymbol{\mu}_-$ are estimated empirically using the training data. One of the limits of this approach is that depending on the dimension d of the problem and the number n of training datapoints, the covariance matrix may not be invertible. In practice, we take the inverse of the matrix $\tilde{\boldsymbol{\Sigma}} = \boldsymbol{\Sigma} + \lambda\mathbf{I}_d$ where $\mathbf{I}_d$ is the identity matrix of dimension d. It can be shown that taking the inverse of $\tilde{\boldsymbol{\Sigma}}$ amounts to performing a quadratic regularization, which may be interpreted as assuming an *a priori* Gaussian distribution for the vector $\mathbf{w}$ with an assumed variance of $\sigma = 2/\sqrt{\lambda}$. Note that LDA is a special case of Fisher discriminant analysis (FDA), which is also widely used for BCIs [WAN 04]. The reader can refer to [HAS 01, Chapter 4] for more details on LDA and its quadratic extension Quadratic Discriminant Analysis

(QDA) when the covariances of each class are not equal. There exist many other extensions of this approach; for example, stepwise LDA allows features to be chosen using statistical tests [KRU 08]. The LDA method may also be extended to work for multiclass classification.

The LDA approach, which is widely used for brain–computer interfaces [BOS 04, BLA 04], is relatively simple to implement and does not involve any hyperparameters (in its non-regularized version; in the regularized version, the λ parameter is added). Among the 18 datasets used in the last BCI competitions [BLA 04, BLA 06, TAN 12], LDA and FDA were used in nine of the methods with the best performance in the classification category.

Logistic regression, a classification method that is also used for BCIs [TOM 07], introduces a logistic cost function (Table 9.1). Although this technique is less widely used than LDA or SVM, it performs excellently for linear classifications, particularly at higher dimensions.

9.1.3.2. *Support Vector Machines (SVM)*

Support Vector Machines or SVMs are obtained by minimizing a criterion of type equation [9.3], where the empirical risk term is the hinge loss (Table 9.1 and Figure 9.2). The most common regularization term is the ℓ_2 regularization term, which effectively maximizes the *margin*, i.e. the minimum distance between the points and the separating hyperplane.

Linear SVMs are a widely used category of classifiers for BCIs [RAK 08, KAP 04]; they have produced state-of-the-art results in detecting event-related potentials [RAK 08, LAB 10] and motor imagery [SCH 05a]. Specifically, SVMs were the best performers in six of 18 of the datasets of the three latest BCI competitions, in particular for the most recent datasets [BLA 04, BLA 06, TAN 12].

Note that SVMs may be extended to train nonlinear classifiers using the *kernel trick*, which replaces the scalar product with a similarity function and thus generates a more complex separating boundary. However, despite some very encouraging results in competitions [LAB 10], using nonlinear classifiers is often viewed as unnecessary for BCIs.

9.2. Specific training methods

In the context of brain–machine interfaces, the objective of methods of statistical learning is essentially to train the function describing the relationship between EEG signals and specific mental states (presence/absence of P300, predefined intentions of movement). However, more advanced, specific methods have also been developed to improve the performance of the process of decoding these mental states, but also to reduce the time required to calibrate the BCI system for new users. In the next few sections, we will present a selection of recent methods that address these key obstacles.

9.2.1. *Selection of variables and sensors*

For certain BCI paradigms such as BCI P300 spellers or BCI motor imagery, *a priori* neurophysiological knowledge allows us to place EEG sensors so that the quality of the recorded signal is nearly optimal. Nevertheless, in certain contexts, such as during the development of new BCI paradigms, or when the regions of interest in the cortex are damaged, it may be desirable to explicitly optimize the positioning of the sensors. This optimization might also be motivated by desire to improve performance by removing sensors that only provide marginal information. The literature on the topic of variable and sensor selection is very rich. We will only attempt a limited review of these methods; we encourage interested readers to refer to the references for more detail [GUY 03]. For BCIs, among the various different available methods, two particular strategies of sensor selection have been studied in most depth.

The first strategy finds the best-performing subset of sensors for recognizing mental states by successively eliminating sensors [LAL 04, SCH 05b, RAK 08]. The principle of the technique is to define a selection criterion (typically the margin of a SVM classifier or an estimate of generalized performance), denoted as C_r. The procedure begins by selecting all sensors and then eliminating one sensor per iteration. The elimination criterion is the following: at each iteration, the performance C_r is calculated for all remaining sensors, followed by the performance C_r^{-j} where the jth sensor has been omitted. The sensor chosen for elimination is the sensor that minimizes:

$$|C_r - C_r^{-j}|$$

which is the minimum performance loss from eliminating a sensor. The stopping criterion is a predefined number of sensors, or too great a decrease in C_r. In computational terms, this elimination criterion may be very expensive to evaluate, however, in certain cases (margin criterion and linear SVM), it has an analytical expression that is easy to calculate [SCH 05b].

Another possible way to select variables and sensors in BCI systems is the technique of applying this selection process while the decision function is being trained. This can be achieved by choosing directly in equation [9.3] a regularization term that induces sparsity in the vector **w** and that implicitly selects variables or sensors. Bach et al. present a review of recent work in this area [BAC 12]. In the context of brain–machine interfaces, the most commonly used regularization terms are based on non-differentiable norms as follows:

– the norm $\ell_1(\mathbf{w}) = \sum_{i=1}^{d} |w_i|$. This norm generates unstructured sparsity in the variables comprising the vector **w**. It is better suited for the selection of variables than for the selection of sensors;

– the mixed norm $\ell_{1,p}(\mathbf{w}) = \sum_{j=1}^{|\{G_j\}|} \left(\sum_{i \in G_j} |w_i|^p\right)^{1/p}$. Here, the sets $\{G_j\}$ form a disjointed partition of the indices 1 to *d*, so that the mixed norm is obtained by calculating the ℓ_1 norm of the vector of dimension $|\{G_j\}|$ whose jth component is given by the ℓ_p norm of the vector of elements indexed by G_j. This norm tends to induce grouped sparsity in the indices contained in a certain subset of the G_j. Thus, if the sets G_j are constructed so that they only contain references to variables linked to a specific sensor, using a regularization term of this type implicitly induces a process of channel selection.

Mixed norms were used for selecting groups of channels for the P300 speller BCI [TOM 10, FLA 14a] and the localization of EEG sources [STR 14]. Using more complexly structured representations such as kernel methods, Jrad *et al.* performed implicit sensor selection in multiple problems of BCI learning using only the ℓ_1 norm [JRA 11]. Although we have only discussed two types of norm here, other regularization terms may be constructed from *a priori* knowledge of the problem at hand, depending on the types of model that we wish to induce [BAC 12].

9.2.2. *Multisubject learning, information transfer*

One of the major hurdles for the widespread application of brain–machine interfaces is the current need for dedicated calibration specific to the person using the interface. Even today, in most cases, before any such interface can be effectively used, a session of EEG signal acquisition must first be organized in order to obtain training data. There are multiple avenues of research based on techniques of statistical learning that might possibly allow this hurdle to be overcome.

One of the possible approaches for building brain–machine interfaces that require less calibration with new users is to use training techniques based on information transfer, or multitask training techniques. These techniques attempt to train multiple decision functions simultaneously and share the set of available data. Thus, for brain–machine interfaces, this is equivalent to multisubject training. The objective of this approach is to attempt to compensate for the fact that only limited training data are available for a given classifier by transferring information from data that are available to other classifiers. This idea has been successfully implemented in several research projects. For example, Devlaminck *et al.* [DEV 11] used this principle to train spatial filters to adapt to subjects with very little available training data. Based on the same principles of multitask training, Alamgir *et al.* [ALA 10] suggest training EEG classifiers for different subjects by modeling their $\mathbf{w}$ vector as the realization of a random variable from a normal distribution. With this model, the average $\mathbf{w}$ vector is used for new subjects, and is subsequently updated as more data becomes available. Remaining within this multitask context, it is possible to select sensors common to multiple subjects using an appropriate mixed norm [FLA 14a]. Experiments show that when training a classifier to recognize P300 signals, significant performance gains are achieved for subjects that otherwise perform poorly without information transfer.

9.3. Performance metrics

This section presents a selection of the performance metrics used to evaluate and compare the advantages of different hypotheses or algorithms, considering the cases of classification and regression separately. These performance metrics, which must always be estimated from data distinct from the training data (see section 9.4 [BLA 06, BLA 04]), test the generalization capacity of the trained hypotheses. Readers can refer to Schlögl et al. [SCH 07] for a more complete treatment of the topic of performance metrics.

9.3.1. *Classification performance metrics*

The traditional performance metrics used for classification problems are based on the confusion matrix $\mathbf{C}$, which for each pair of classes (i, j) records the number $C_{i,j}$ of instances of class i that have been assigned to class j. If the classification is perfect, the matrix $\mathbf{C}$ is diagonal. The accuracy (ACC) is the fraction of properly classified instances (which are on the diagonal of the confusion matrix), which is equal to the $0 - 1$ error (section 9.1.2) and estimates the Bayes error rate of the classifier. This rate was used in most BCI competitions to evaluate the performance of the motor imagery classifiers [BLA 06, BLA 04].

However, the accuracy is not a good measure of performance in the case of strongly unequal class sizes: indeed, if one class contains 99% of the datapoints, the trivial classifier that groups all points into the same class has an accuracy of 99%. An alternate, more appropriate metric is Cohen's Kappa coefficient [CAR 96], which was also used in BCI competitions [BLA 06, BLA 04], defined by

$$\kappa = \frac{ACC - p_e}{1 - p_e}, \quad \text{with} \quad p_e = \frac{1}{N^2} \sum_i (\sum_j C_{i,j})(\sum_j C_{j,i})$$

where p_e is the probability that a random classification is correct. The mutual information is another performance metric that is often used for BCIs [NYK 01]. It is based on information theory [BLA 06, SCH 07] and measures in bits the rate of information transmitted to the machine by the interface.

Another metric that is particularly well-suited for classes of unequal sizes is the area under the receiver operating characteristics (ROC) curve (AUC), or the Mann–Whitney–Wilcoxon score, given by:

$$AUC(f(\cdot)) = Pr(f(\mathbf{x}) > f(\mathbf{x}')|y > y')$$

Consider a binary classification hypothesis $f(\cdot)$ that takes values in $\mathbb{R}$. For any threshold $\tau \in \mathbb{R}$, we can define the classifier f_τ that classes $\mathbf{x}$ in the positive class iff $f(\mathbf{x}) > \tau$, as well as the true positive rate TPR_τ $(P(f(\mathbf{x}) > \tau|y = 1))$ and the false positive rate FPR $_\tau$ $(P(f(\mathbf{x}) < \tau|y = 1))$. The monotone curve of the points (FPR_τ, TPR_τ) is called the ROC curve (Figure 9.3). If there exists a threshold such that the classifier $f(\cdot)_\tau$ is perfect, the ROC curve passes through (0,1) (0% false negatives and 100% true positives). If instead the ROC curve coincides with the diagonal, any improvement in the true positive rate is offset by an equal decrease in the false positive rate: in other words, the classifier provides no information. More generally, the area under the ROC curve measures the quality of the hypothesis $f(\cdot)$, regardless of whether the classes are of similar size (since the coordinates FPR_τ and TPR_τ are percentages).

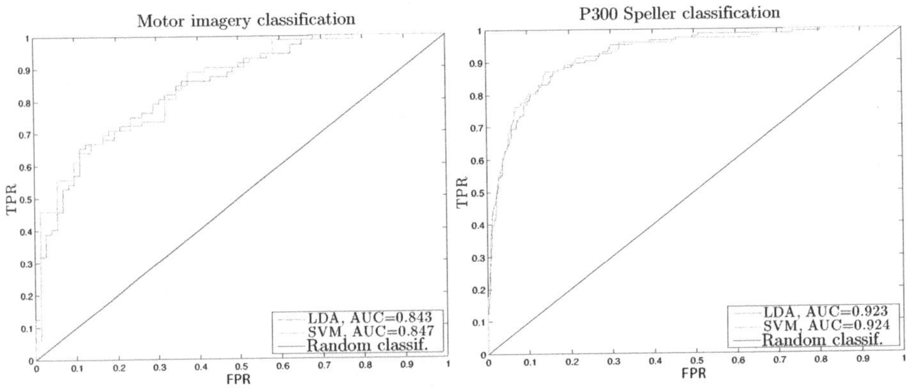

Figure 9.3. *ROC curves for an application in motor imagery (left) and P300 speller (right). Performance metric AUC: area under the ROC curve. Horizontal axis, false positive rate (FPR); vertical axis, true positive rate (TPR). For a color version of the figure, see www.iste.co.uk/clerc/interfaces1.zip*

The AUC criterion was used in the BCI MLSP 2010 competition [HIL 10]; for the P300 speller in particular, since by construction, the positive class contains far fewer instances than the negative class.

9.3.2. *Regression performance metrics*

For regression tasks, such as predicting the motion of limbs, other performance metrics need to be considered [PIS 08]. Similarly to classification, the performance metrics are often linked to the cost term of the error, which is minimized during training. Thus, for least squares regression, the performance metric is the mean square error or ℓ_2 cost (Table 9.1). The disadvantage of this metric is that it is not normalized (it depends on the amplitude of the data y_i), which makes comparing performance difficult for tasks such as predicting finger flexion [LIA 12] and motion through space [PIS 08].

One possible alternative is to take the correlation between the predicted value $f(\mathbf{x})$ and the observed value y as a performance metric. This metric is equal to 1 in the case of monotone prediction ($f(\mathbf{x})$ increases with y); if the hypothesis provides zero information, the correlation is equal to 0. However, correlation does not take into account the true value of the prediction, which is important for tasks such as controlling the position of a cursor [WU 06].

9.4. Validation and model selection

Validating the results obtained in a given application serves two purposes in statistical learning: evaluating the chosen performance metric (section 9.3) and optimizing the hyperparameters of the algorithm. Readers can refer to [DUD 99, HAS 01, JAP 14] for a more in-depth treatment of the topic of validation.

9.4.1. *Estimation of the performance metric*

As mentioned earlier, the performance of a classifier $f(\cdot)$ on the training data is ultimately not what we are interested in; by increasing the complexity of $f(\cdot)$, it is always possible to construct a classifier that performs perfectly on the known data. The end-goal is the capacity of generalization of $f(\cdot)$, that is to say its performance on future data. We can attempt to evaluate the general performance of $f(\cdot)$ by partitioning the available data into training data, which is used to optimize $f(\cdot)$, and test data, which is used to estimate the performance of $f(\cdot)$ in general. In doing so, we assume that the test data extracted from the available dataset is a good representation of the future data that the classifier will need to predict.

If there is plenty of available data, estimating the performance metric is not particularly difficult: there are sufficient data to train a classifier $f(\cdot)$, and there are sufficient (other) data for estimating the performance of $f(\cdot)$. But this task becomes more complex when there are limited available data, which is the case for BCIs due to the cost of data acquisition. In this context, we must find the right balance between reducing the quantity of training data (and therefore reducing the quality of the training hypothesis) and reducing the quantity of test data (and therefore reducing the reliability of the quality estimate).

In practice, there are three main approaches to this problem (Figure 9.4). The preferred approach depends on the way that the training data and the test data are constructed.

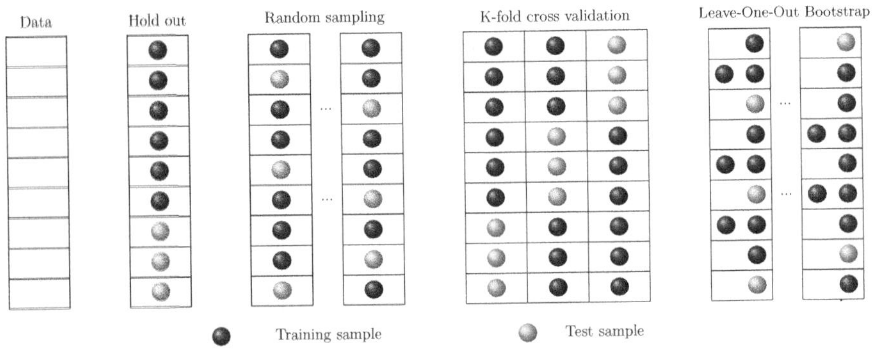

Figure 9.4. *Illustration of the different data partitions presented in this chapter used to estimate the generalization error*

9.4.1.1. *Random sampling*

The process known as the *hold-out* method randomly partitions the n datapoints into training data $\mathcal{E}$ and test data $\mathcal{T}$, uses $\mathcal{E}$ to train a hypothesis $f(\cdot)$ and then measures the performance of the hypothesis $f(\cdot)$ on $\mathcal{T}$. This performance might potentially have a large variance (if the datapoints in $\mathcal{E}$ are easy to classify, and those in $\mathcal{T}$ are difficult, or vice versa). To reduce the variance, the method of *random sampling* repeats the hold-out method by considering K independent samplings for the training data and the test data. The performances obtained on each of the different sets of test data are then averaged. The average performances and their confidence intervals for two different classifiers may be used as a basis for comparison to determine whether one is superior to the other to a certain degree of significance using a standard hypothesis test. Since training is repeated K times on independent sets of training data, and the performances on the test data are averaged, this process makes the performance metric more robust against the randomness inherent in sampling at the cost of increasing the number of calculations by a factor of K.

9.4.1.2. *K-cross-validation and leave-one-out*

K-fold cross-validation partitions the n datapoints into K subsets of equal size. A series of hold-out processes is then performed by taking each of the subsets as the testing dataset, and the remaining data as the training dataset. As above, the performance is estimated by the average of the performances on the K testing subsets. Compared to random sampling, cross-validation reduces the variance of the estimate (each datapoint is included exactly once in the testing dataset). When the number of available datapoints is very low, we may in particular choose K equal to the number n of points: cross-validation is then equivalent to the *leave-one-out* (LOO) method, which performs training on all datapoints except one, saving the final point to test the trained classifier.

The choice of K involves establishing a compromise between the bias and the variance: as K increases, the bias of the estimate decreases (given some weak assumptions [HAS 01]) but the variance increases, because the various training datasets are strongly correlated. The complexity also increases with K. Finally, the choice of K fundamentally depends on the number n of datapoints. If n is large relative to the number d of descriptive attributes, low values of K may be considered sufficient ($K = 2 \dots 10$), and the variance may be reduced by repeating cross-validation on five different partitions of the data.

For BCIs, cross-validation is normally used when the performance is good (see, for example, Labbé *et al.* [LAB 10] on estimating the performance of processes for recognizing P300 signals).

9.4.1.3. *Bootstrapping*

Bootstrapping constructs the training data using n uniform samples *with replacement* from the n available datapoints. Certain points can therefore be present multiple times in the training data, and others may be omitted; on average, the fraction of non-selected points is 37% of the available data. Classifier performance estimation is performed on the whole of the available dataset; as above, the performance is averaged over multiple different samples for the training data. This method of estimation is less optimistic than LOO; the bias is similar to the bias of cross-validation with $K = 2$ [HAS 01]. Despite this bias, performance rankings of hypotheses using bootstrapping are usually considered reliable.

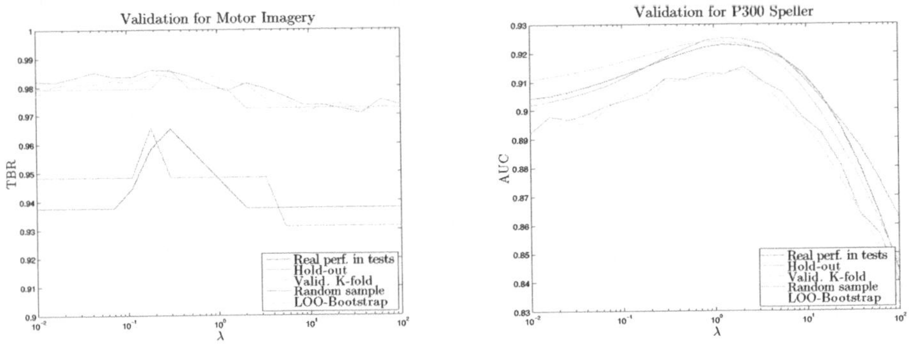

Figure 9.5. *Illustration of the different methods of estimating the generalization error on two different classification datasets. Left: Motor imagery classification with very few datapoints, performance metric = accuracy (ACC). Right: Detection of triggered potential for P300 speller, performance metric = area under the ROC curve (AUC). For a color version of the figure, see www.iste.co.uk/clerc/interfaces1.zip*

9.4.2. *Optimization of hyperparameters*

Training algorithms often involve hyperparameters, such as the weight of the regularization term (equation [9.3], λ parameter), or the kernel parameters of a nonlinear SVM (standard deviation of a Gaussian kernel, degree of a polynomial kernel). Naturally, it is desirable to find the values of hyperparameters that produce the optimal performance in generalization. The recommended procedure for doing so is as follows. The available data are partitioned into three sets: the training set $\mathcal{E}$, the test set $\mathcal{T}$ and the validation set $\mathcal{V}$.

Starting with $\mathcal{E}$ and a vector θ of hyperparameters, the performance is measured on the whole of the test set $\mathcal{T}$, denoted as $\mathcal{F}(\theta)$. The question may then be rephrased as finding the optimal vector of hyperparameters θ^*:

$$\theta^* = \arg\max\{\mathcal{F}(\theta)\}$$

After finding θ^* (see below), it is useful to recalculate the value of the performance metric. Indeed, since θ^* was determined using the information present in $\mathcal{E}$ and $\mathcal{T}$, the performance metric $\mathcal{F}(\theta)$ is optimistic on $\mathcal{T}$[2]. So, θ^* is used for training on $\mathcal{E} \bigcup \mathcal{T}$, and the classifier thus obtained is evaluated on the validation set $\mathcal{V}$, which has not yet been considered.

θ^* may be determined in one of three ways. If the algorithm involves a low number of hyperparameters ($\theta \in \mathbb{R}^d$, with $d = 2.3$), the usual method is an exhaustive search over a grid in the parameter space [CHA 11, RAK 08]. This grid may be regular (e.g. the degree of a polynomial taken within 1, 2,...,10) or not (e.g. varying the regularization weight by powers of 10, $\lambda = 10^i$, for $i = -3 \ldots 3$) . An example of exhaustive search for $\theta*$ is shown in Figure 9.5. The advantage of this method is its simplicity; however, the cost increases exponentially with the number of hyperparameters involved in the algorithm.

If the performance metric $\mathcal{F}$ (or an approximation or a bound for this metric) is differentiable, $\theta*$ may be found using gradient descent methods. This approach was used, for example, to optimize the hyperparameters of a SVM [CHA 02, FLA 14b], and for linear discriminant analysis with two classes [KEE 06]. Although complex to implement, this method has the advantage of being less costly than the previous method when the number of parameters is large.

Finally, if $\mathcal{F}$ is known only as a black box (an algorithm that returns the value of $\mathcal{F}(\theta)$ for each θ), we can use stochastic methods or black box optimization, ranging from simulated annealing to evolution-based strategies [HAN 01]. In general, stochastic optimization algorithms apply a dynamic distribution to the search space, which is gradually biased toward the regions of best performance. The *covariance–matrix–adaption* algorithm[3] achieves this by adapting the parameters of a Gaussian distribution so that it gradually converges toward an optimum of the target function. This method is more costly than the previous methods: it requires $\mathcal{F}(\theta)$ to be evaluated for a relatively large number of hyperparameter vectors θ, whereby calculating one single $\mathcal{F}(\theta)$ requires training and evaluating a classifier. Hybrid methods of optimization combining stochastic optimization of $\mathcal{F}$ and the training of an approximation of $\mathcal{F}$, called a surrogate model (*surrogate model-based*

2 It would be cheating to assume that $\mathcal{F}(\theta^*)$ represents the true performance of the classifier.

3 https://www.lri.fr/ hansen/cmaes_inmatlab.html.

optimization), are therefore used for optimizing the hyperparameters of general learning algorithms [BER 12, BAR 13], and, in the particular context of BCIs, for the selection of variables [GAR 03, SCH 03, COR 11].

9.5. Conclusions

This chapter gives a short presentation of the principal methods of supervised learning used with brain–computer interfaces, with a focus on the practical challenges posed by their implementation. References to some recent and promising work were included, so that readers may explore these topics in more depth according to their individual context. Finally, the validation of obtained results is an essential aspect of machine learning, both for optimizing the implementation of an algorithm and as a basis for a rigorous comparison of different algorithms.

9.6. Bibliography

[ALA 10] ALAMGIR M., GROSSE-WENTRUP M., ALTUN Y., "Multitask learning for brain–computer interfaces", *Proceedings of the 13th International Conference on Artificial Intelligence and Statistics*, pp. 17–24, 2010.

[BAC 12] BACH F., JENATTON R., MAIRAL J. *et al.*, "Optimization with sparsity-inducing penalties", *Foundations and Trends® in Machine Learning*, vol. 4, no. 1, pp. 1–106, 2012.

[BAR 13] BARDENET R., BRENDEL M., KÉGL B. *et al.*, "Collaborative hyperparameter tuning", *International Conference on Machine Learning (ICML)*, vol. 28, pp. 199–207, 2013.

[BEN 10] BÉNAR C.B., CHAUVIÈRE L., BARTOLOMEI F. *et al.*, "Pitfalls of high-pass filtering for detecting epileptic oscillations: a technical note on "false" ripples", *Clinical Neurophysiology*, vol. 121, pp. 301–310, 2010.

[BER 12] BERGSTRA J., BENGIO Y., "Random search for hyper-parameter optimization", *The Journal of Machine Learning Research*, vol. 13, no. 1, pp. 281–305, 2012.

[BLA 04] BLANKERTZ B., MÜLLER K., CURIO G. *et al.*, "The BCI competition 2003: progress and perspectives in detection and discrimination of EEG single trials", *IEEE Transactions on Biomedical Engineering*, vol. 51, no. 6, pp. 1044–1051, 2004.

[BLA 06] BLANKERTZ B., MÜLLER K., KRUSIENSKI D.J. *et al.*, "The BCI competition III: validating alternative approaches to actual BCI problems", *IEEE Transactions on Neural Systems and Rehabilitation Engineering* vol. 14, no. 2, pp. 153–159, 2006.

[BOS 04] BOSTANOV V., "BCI competition 2003-data sets Ib and IIb: feature extraction from event-related brain potentials with the continuous wavelet transform and the t-value scalogram", *IEEE Transactions on Biomedical Engineering*, vol. 51, no. 6, pp. 1057–1061, 2004.

[CAR 96] CARLETTA J., "Assessing agreement on classification tasks: the kappa statistic", *Computational linguistics*, vol. 22, no. 2, pp. 249–254, 1996.

[CHA 02] CHAPELLE O., VAPNIK V., BOUSQUET O. *et al.*, "Choosing multiple parameters for support vector machines", *Machine Learning*, vol. 46, nos. 1–3, pp. 131–159, 2002.

[CHA 07] CHAPELLE O., "Training a support vector machine in the primal", *Neural Computation*, vol. 19, no. 5, pp. 1155–1178, 2007.

[CHA 11] CHANG C.-C., LIN C.-J., "LIBSVM: a library for support vector machines", *ACM Transactions on Intelligent Systems and Technology (TIST)*, vol. 2, no. 3, p. 27, 2011.

[COR 11] CORRALEJO R., HORNERO R., ALVAREZ D., "Feature selection using a genetic algorithm in a motor imagery-based brain computer interface", *Annual International Conference of the IEEE on Engineering in Medicine and Biology Society (EMBC)*, pp. 7703–7706, 2011.

[DEV 11] DEVLAMINCK D., WYNS B., GROSSE-WENTRUP M. *et al.*, "Multisubject learning for common spatial patterns in motor-imagery BCI", *Computational Intelligence and Neuroscience*, vol. 2011, p. 8, 2011.

[DIE 98] DIETTERICH T., "Approximate statistical tests for comparing supervised classification learning algorithms", *Neural Computation*, vol. 10, no. 1, pp. 1895–1923, 1998.

[DRA 81] DRAPER N., SMITH H., *Applied Regression Analysis*, John Wiley, New York, 1981.

[DUD 99] DUDA R.O., HART P.E., STORK D.G., *Pattern Classification*, John Wiley & Sons, Hoboken, NJ, 1999.

[FLA 14a] FLAMARY R., JRAD N., PHLYPO R. *et al.*, "Mixed-norm regularization for brain decoding", *Computational and Mathematical Methods in Medicine*, Hindhawi Publishing, Cairo, Egypt, vol. 2014, no. 1, pp. 1–13, 2014.

[FLA 14b] FLAMARY R., RAKOTOMAMONJY A., GASSO G., "Learning constrained task similarities in graph-regularized multi-task learning", in SUYKENS J.A.K. , SIGNORETTO M., ARGYRIOU A. (eds), *Regularization, Optimization, Kernels, and Support Vector Machines*, Chapman and Hall/CRC, 2014.

[GAR 03] GARRETT D., PETERSON D.A., ANDERSON C.W. *et al.*, "Comparison of linear, nonlinear, and feature selection methods for EEG signal classification", *IEEE Transactions on Neural Systems and Rehabilitation Engineering*, vol. 11, no. 2, pp. 141–144, 2003.

[GUY 03] GUYON I., ELISSEEFF A., "An introduction to variable and feature selection", *The Journal of Machine Learning Research*, vol. 3, pp. 1157–1182, 2003.

[HAN 01] HANSEN N., OSTERMEIER A., "Completely derandomized self-adaptation in evolution strategies", *Evolutionary Computation*, vol. 9, no. 2, pp. 159–195, 2001.

[HAS 01] HASTIE T., FRIEDMAN J., TIBSHIRANI R., *The Elements of Statistical Learning*, Springer, Berlin, 2001.

[HIL 10] HILD K.E., KURIMO M., CALHOUN V.D., "The sixth annual MLSP competition", *IEEE International Workshop on Machine Learning for Signal Processing (MLSP)*, pp. 107–111, 2010.

[JAP 14] JAPKOWICZ N., SHAH M., *Evaluating Learning Algorithms A Classification Perspective*, Cambridge University Press, Cambridge, 2014.

[JRA 11] JRAD N., CONGEDO M., PHLYPO R. *et al.*, "sw-SVM: sensor weighting support vector machines for EEG-based brain–computer interfaces", *Journal of Neural Engineering*, vol. 8, no. 5, p. 056004, 2011.

[KAP 04] KAPER M., MEINICKE P., GROSSEKATHOEFER U. *et al.*, "BCI competition 2003-data set IIb: support vector machines for the P300 speller paradigm", *IEEE Transactions on Biomedical Engineering*, vol. 51, no. 6, pp. 1073–1076, 2004.

[KEE 06] KEERTHI S.S., SINDHWANI V., CHAPELLE O., "An efficient method for gradient-based adaptation of hyperparameters in SVM models", *Advances in Neural Information Processing Systems*, vol. 19, pp. 673–680, 2006.

[KRU 08] KRUSIENSKI D.J., SELLERS E.W., MCFARLAND D.J. *et al.*, "Toward enhanced P300 speller performance", *Journal of Neuroscience Methods*, vol. 167, no. 1, pp. 15–21, 2008.

[LAB 10] LABBÉ B., TIAN X., RAKOTOMAMONJY A., "MLSP competition, description of third place method,", *IEEE International Workshop on Machine Learning for Signal Processing (MLSP)*, pp. 116–117, 2010.

[LAL 04] LAL T.N., SCHRODER M., HINTERBERGER T. *et al.*, "Support vector channel selection in BCI", *IEEE Transactions on Biomedical Engineering*, vol. 51, no. 6, pp. 1003–1010, 2004.

[LIA 12] LIANG N., BOUGRAIN L., "Decoding finger flexion from band-specific ECoG signals in humans", *Frontiers in Neuroscience*, vol. 6, no. 91, pp. 1–6, 2012.

[LOT 07] LOTTE F., CONGEDO M., LÉCUYER A. *et al.*, "A review of classification algorithms for EEG-based brain–computer interfaces", *Journal of Neural Engineering*, vol. 4, 2007.

[LOT 11] LOTTE F., GUAN C., "Regularizing common spatial patterns to improve BCI designs: unified theory and new algorithms", *IEEE Transactions on Biomedical Engineering*, vol. 58, no. 2, pp. 355–362, 2011.

[NYK 01] NYKOPP T., Statistical modeling issues for the adaptive brain interface, Master's Thesis, Helsinki University of Technology, 2001.

[PIS 08] PISTOHL T., BALL T., SCHULZE-BONHAGE A. *et al.*, "Prediction of arm movement trajectories from ECoG-recordings in humans", *Journal of Neuroscience Methods*, vol. 167, no. 1, pp. 105–114, 2008.

[RAK 08] RAKOTOMAMONJY A., GUIGUE V., "BCI competition III: dataset II - ensemble of SVMs for BCI P300 speller", *IEEE Transactions on Biomedical Engineering*, vol. 55, no. 3, pp. 1147–1154, 2008.

[SCH 03] SCHRODER M., BOGDAN M., HINTERBERGER T. *et al.*, "Automated EEG feature selection for brain computer interfaces", *First International IEEE EMBS Conference on Neural Engineering*, IEEE, Cancun, Mexico, pp. 626–629, 2003.

[SCH 05a] SCHLÖGL A., LEE F. *et al.*, "Characterization of four-class motor imagery EEG data for the BCI-competition", *Journal of Neural Engineering*, vol. 2, no. 4, p. L14, 2005.

[SCH 05b] SCHRÖDER M., LAL T.N., HINTERBERGER T. *et al.*, "Robust EEG channel selection across subjects for brain-computer interfaces", *EURASIP Journal on Applied Signal Processing*, vol. 2005, pp. 3103–3112, 2005.

[SCH 07] SCHLÖGL A., KRONEGG J., HUGGINS J.E. *et al.*, "Evaluation criteria for BCI research", *Toward Brain–computer Interfacing*, The MIT Press, Cambridge, Massachusetts, USA, 2007.

[SMO 04] SMOLA A.J., SCHÖLKOPF B., "A tutorial on support vector regression", *Statistics and Computing*, vol. 14, no. 3, pp. 199–222, 2004.

[STR 14] STROHMEIER D., HAUEISEN J., GRAMFORT A., "Improved MEG/EEG source localization with reweighted mixed-norms", *International Workshop on Pattern Recognition in Neuroimaging*, IEEE, Tubingen, Allemagne, pp. 1–4, 2014.

[TAN 12] TANGERMANN M., MÜLLER K.-R., AERTSEN A. *et al.*, "Review of the BCI competition IV", *Frontiers in Neuroscience*, vol. 6, 2012.

[TOM 07] TOMIOKA R., AIHARA K., MÜLLER K.-R., "Logistic regression for single trial EEG classification", *Advances in Neural Information Processing Systems*, vol. 19, pp. 1377–1384, 2007.

[TOM 10] TOMIOKA R., MÜLLER K., "A regularized discriminative framework for EEG analysis with application to brain–computer interface", *NeuroImage*, vol. 49, no. 1, pp. 415–432, 2010.

[VAP 98] VAPNIK V., *Statistical Learning Theory*, Wiley, Hoboken, NJ, 1998.

[WAN 04] WANG Y., ZHANG Z., LI Y. *et al.*, "BCI competition 2003-data set IV: an algorithm based on CSSD and FDA for classifying single-trial EEG", *IEEE Transactions on Biomedical Engineering*, vol. 51, no. 6, pp. 1081–1086, 2004.

[WU 06] WU W., GAO Y., BIENENSTOCK E., *et al.*, "Bayesian population decoding of motor cortical activity using a Kalman filter", *Neural Computation*, vol. 18, no. 1, pp. 80–118, 2006.

PART 3

Human Learning and Human–Machine Interaction

10

Adaptive Methods in Machine Learning

The signals used by non-invasive BCIs (EEG, MEG, etc.) fluctuate strongly over time, between different sessions, but also within each session itself, based on the subject's levels of fatigue, motivation or physical changes such as sensor positioning and impedance. The goal of adaptive methods is to tackle the problem of the *non-stationarity* of the signal, or more precisely, the variability of irrelevant information as opposed to the variability of relevant information which allows to control the various degrees of freedom of the interface. These adaptive methods act by modifying the response function during BCI use, in order to maintain (or even improve) the performance of the interface.

10.1. The primary sources of variability

Human biomedical research distinguishes between two principal types of variability, particularly in the domain of cognitive neurosciences and neuroimagery: intrasubject variability (within single subjects) and intersubject variability (between subjects) [FRI 02]. BCIs are affected by both types of variability. Indeed, the goal of a BCI is to optimize the interaction afforded to a given individual. Especially in medical contexts, the success of the BCI is largely dependent on its capacity to adapt to each patient, depending on the medical circumstances of the individual. In order to be

Chapter written by Maureen CLERC, Emmanuel DAUCÉ and Jérémie MATTOUT.

applicable to as many patients as possible, BCIs must be capable of adapting to intersubject differences. Furthermore, to guarantee high levels of stability and performance, BCIs must be robust against the signal variability that occurs at individual scales. This last point is particularly crucial for BCIs that, by definition, operate online; these BCIs must rapidly establish a link between small numbers of observations, corrupted by noise, and machine decisions.

This section offers a brief review of the primary factors that govern and influence these two types of variability.

10.1.1. *Intrasubject variability*

In essence, there are two major sources of intrasubject variability: technical factors and human factors. Both can affect the robustness of a BCI, during usage, or from one session to the next.

Technical factors include anything that does not depend on the subject but that might potentially alter the reproducibility of the relevant features of the recorded signals. It is difficult to guarantee identical recording conditions over multiple sessions (positioning and impedance of the electrodes for EEG; development of fibrosis during invasive recordings). Within the same session, the signal from one of the EEG sensors can suddenly begin to deteriorate, either due to poor contact (insufficient gel, electrode slipping or becoming detached) or due to the sensor developing a spontaneous defect. In particular, if the impedance or the signal-to-noise ratio of the reference electrode changes, it will affect all of the signals (see [CLE 16], sections 8.2 and 9.3). If a BCI uses non-adaptive algorithms whose parameters are trained during a calibration phase and then remain unchanged for subsequent usage, these kinds of events can have a dramatic impact on the performance. Imagine, for example, a spatial filter with a high weight on one particular sensor (see Chapter 7), which then suddenly loses signal: the BCI is immediately rendered ineffective. If we wish to prevent degraded performance, given that most BCIs are intended for repeated use over multiple sessions for extended periods of time, the use of adaptive methods appears unavoidable.

The *human factors* are the changes in mental state that might alter the physiological markers on which the interaction is based. The markers of interest such as event-related responses and spontaneous rhythms are affected by many of these factors, including for example fluctuations in vigilance, decreased motivation, moments of inattention or even habituation [KOH 07], which also equally affect learning. The goal is to eliminate the negative effects of these fluctuations, or alternatively to be able to measure and interpret them for the opportunity of exploiting them [BER 07]. However, the neurophysiological correlates of attention levels, fatigue and even learning are still poorly understood; they are currently the object of their own lines of research in cognitive neuroscience. This research specifically aims to improve our understanding of the origins of the intrasubject variability in behavioral performance [WEI 06, DRE 11, BOM 15]. In future, the results of this research may greatly enhance the development of more robust, adaptive BCIs. Finally, intentions rather than objects some differences in cerebral activity are rooted in the context, which defines the objects and the expectations of the user. The user context (at hospital, at home, alone or accompanied) plays an important role in this, but so does the experimental context, such as the fact of being online with a working BCI, or offline during a calibration phase. This phenomenon can limit the capacity of generalization of the calibration data, and provides another reason for choosing to implement adaptive methods.

10.1.2. *Intersubject variability*

More difficult to tackle than intrasubject variability, intersubject variability is another one of the major challenges of BCIs. Ideally, while remaining robust against variability at individual scales, BCIs should build upon working principles and neurophysiological markers that are applicable to the majority of subjects or patients.

Intersubject variability can be anatomical in origin, for example occurring due to differences in the gyrification of the cortex. Since EEG is very sensitive to the orientation of the cortical generators, sources of cerebral activity that are close to each other but oriented in different ways can produce very different measurements at the electrodes. Intersubject variability has been identified in a number of cortical structures, for example the sensorimotor cortex [GRA 91] and even the medial cingulate cortex, which is composed of either one or two sulci depending on the subject [AMI 13]. The medial cingulate cortex is the primary location of markers known as error potentials (see section 5.5). The intersubject variability of these markers is particularly significant in the context of BCIs.

Intersubject variability may also have a functional origin; the responses to a stimulus differ from subject to subject, all the more so that their latency is long: the variability of P300 waves is much higher than that of the spikes of early auditory responses [DAL 97]. Similarly, there is intersubject variability in the frequency of spontaneous rhythms. For instance in MEG, the posterior alpha activity (occipital and parietal) was measured at 10.3 Hz with an intersubject standard deviation of 2.8 Hz [HAE 14]. These differences are the reason behind the standard practice of starting with a calibration session for each new user, for BCIs exploiting event-related responses as well as oscillatory activities. The calibration session optimizes physiological markers for each individual by characterizing their temporal, spectral and spatial features (see section 7.2). Transfer learning approaches aim to eliminate the necessity of calibration session by initializing the BCI with a database of other subjects or patients (see section 9.2.2). This process is naturally limited by intersubject variability, but its implementation alongside adaptive methods (explained below, section 10.3.2) is a promising prospect for optimizing both usage, comfort and BCI performance.

Intersubject differences can also arise due to psychological factors, personality traits or simply age differences. These aspects will need to be accounted for when designing BCIs for widespread usage.

Finally, although currently BCIs are mostly being tested and evaluated with healthy subjects, we should not forget that there can be important differences in cerebral activity between healthy subjects and the patients who are the target population of this technology. Moreover, interpatient variability is possibly even stronger than intersubject variability within the general population, due to differences in the etiology and the evolution of the patients' pathologies.

For example, the event-related auditory responses of patients in a coma or waking from a coma can be very atypical, very variable, weaker and delayed or even absent, depending on neurological impairments and the cause of the coma (trauma or anoxia), and also depending on the evolution of the state of the patient over time (see [MOR 14] and Chapter 1 in Volume 2 [CLE 16]).

These parameters must be taken into consideration when developing and validating BCIs for clinical purposes.

10.2. Adaptation framework for BCIs

Figure 10.1 provides a general framework for studying variability in the context of BCIs. The BCI attempts to interpret the measurements, denoted x, originating from the cerebral activity of a subject with the objective of making a decision. The decision is based on the analysis of certain indicators y of the cerebral state of the subject when he or she receives the stimulus or feedback u.

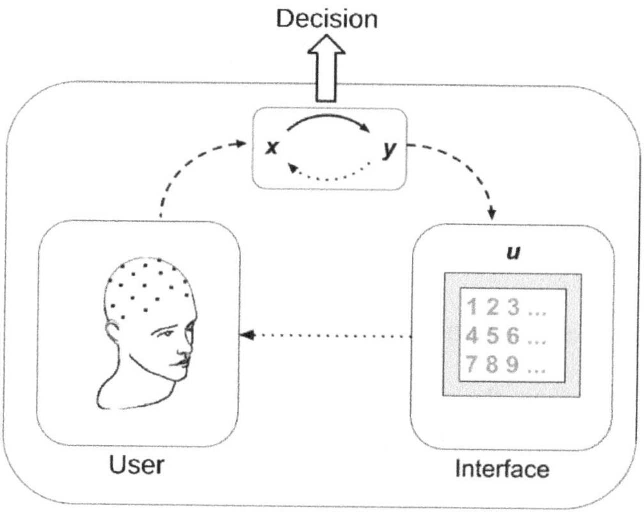

Figure 10.1. *General framework for studying variability in BCIs. The relationship between the cerebral state y and the measurements x is subject to variability, and this must be taken into account when estimating the cerebral states y*

The relationship between the cerebral state y and the measurements x can be expressed in two different ways:

– as a function explicitly linking x and y according to the logic of the statistical decoding of the cerebral activity (classification of this activity). One can then think of y as a label;

– as a generative model describing as realistically as possible the causality between y and x. y is then thought of as a hidden variable representing the user's intention. Decoding is performed by inverting the model (Bayesian approach).

In each of these two approaches, the same objects are sometimes called different names. For the sake of clarity, Table 10.1 lists the links between the different instances of terminology for this chapter. The rest of the chapter is organized in two parts, which presents the two approaches for describing variability: statistical decoding and generative models.

	Statistical decoding	Generative model
$\mathbf{x}$	Data, observation, feature, covariable	Observation, feature
y	Label, variable	State, hidden variable
u	Stimulations	Stimulations, experimental design
f	Prediction function, classifier	
$M(\theta)$		Generative model, potentially dependent on parameters θ

Table 10.1. *Commonly used terminology from two different perspectives: statistical decoding and generative modeling*

10.3. Adaptive statistical decoding

This section adopts the perspective of a decoding process applied to the recorded activity; the goal is to estimate a prediction function f that, given any observation $\mathbf{x}$, returns the predicted label $y = f(\mathbf{x})$.

10.3.1. *Covariate shift*

As explained in section 9.1, in supervised learning, the prediction function is estimated using a dataset whose labels are known. This dataset, known as the training dataset, is usually obtained during a dedicated calibration session. The prediction function may then be used to label new data, known as the test data. Applications of BCIs rely on being able to label the test data online.

Over the course of a single session, the statistical distribution of the data might be subject to change (and *a fortiori* between different sessions). If these changes occur between the training and testing phases, the labels of the new test data risk being incorrect, consequently leading to incorrect decoding (upon which the BCI relies).

This situation may be modeled using the concept of *covariate shift*: the distribution of the observations (or covariates) $\mathbf{x}$ is non-stationary, whereas the conditional distribution $P(y|\mathbf{x})$ on the other hand is stationary (Figure 10.2). Note that the condition that $P(y|\mathbf{x})$ is stationary implies that, regardless of its shift, a covariate $\mathbf{x}$ that returns to a previous state at a later point in time still corresponds to the same label y. Covariate shift models provide reasonable formalisms for BCIs when $\mathbf{x} \in \mathbb{R}^d$ with relatively high d, and when the values of the labels y are discrete (classification).

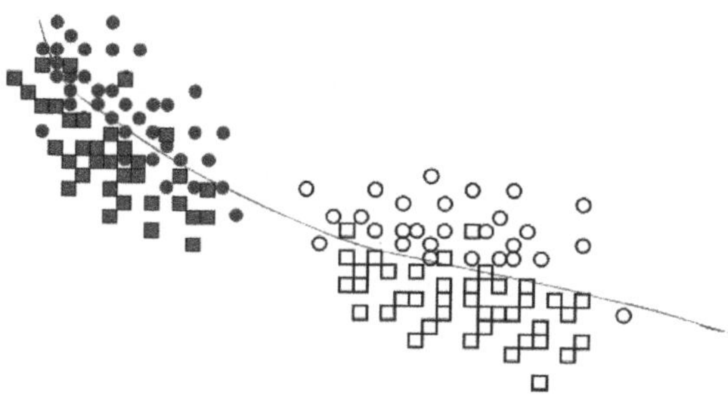

Figure 10.2. *Illustration of covariate shift, in two dimensions ($\mathbf{x} \in \mathbb{R}^2$). Each circle (square) indicates an observation $\mathbf{x}_i$ for task 1 (2). The covariate shift may be seen in the differences between the probability distributions of the colored points (training data) and the white points (test data)*

Estimating the prediction function is traditionally achieved by minimizing an empirical risk term (equation [9.2], Chapter 9) recalled below:

$$R_{emp}(f) = \frac{1}{n}\sum_{i=1}^{n} L(y_i, f(\mathbf{x}_i))$$

In order to compensate for covariate shift, the empirical risk should be weighted by an "importance" function: the ratio (assumed finite) of the probabilities of the test and training data:

$$\frac{p_{test}(\mathbf{x})}{p_{training}(\mathbf{x})}. \quad [10.1]$$

The importance-weighted empirical risk then becomes:

$$R_{\mathcal{I}emp}(f) = \frac{1}{n}\sum_{i=1}^{n} \frac{p_{test}(\mathbf{x}_i)}{p_{training}(\mathbf{x}_i)} L(y_i, f(\mathbf{x}_i)) \quad [10.2]$$

If the importance function [10.1] is not known, it may be estimated, for example using parametric models with Gaussian kernels [SUG 07, LI 10].

10.3.2. *Classifier adaptation*

10.3.2.1. *Sliding window retraining*

In mathematical terms, adaptive learning is an optimization problem in which the current performance must be optimized while preserving the performances acquired during previous training phases. A mathematical analysis in [KIV 04], that, in order to follow the variations of the environment without overloading the storage and computational capacity, it is often necessary to delete the oldest datapoints to make space for more recent information. This approach is also known as sliding window retraining.

Several methods for BCIs have been suggested to allow online data to be leveraged at regular intervals. The most common method is to begin with an initial, generic classifier and then *retrain* this classifier in real time. The generic classifier is calculated from a multisubject database that is supplied beforehand (transfer learning, see section 9.2.2). In principle, this classifier can be used immediately with any new subject, without calibration. However, it can be expected to perform fairly poorly. Adaptive approaches attempt to improve this initial classifier by regularly updating it with information gathered during usage. The method suggested by [VID 11] updates the classifier in three stages. The first two stages use a supervised adaptive method, for which the subject must copy a series of words, and the last stage uses an unsupervised adaptive method, during which the subject uses the interface freely. In contrast, the approaches suggested by [LI 06, KIN 12a], inspired by the EM algorithm [DEM 77], are entirely unsupervised. A probabilistic model is constructed from an initial, multisubject dataset. The data acquired during usage are gradually added to the initial dataset to refine the model, *without any labeling information*. The simulations in [KIN 12b] show that in most cases the algorithm is capable of constructing useful classifiers, even without any initial data.

10.3.2.2. *Gradient descent*

Another family of adaptive methods is the family of "stochastic gradient methods". These methods were suggested at a very early stage for neural networks. The approach assumes that it is possible to estimate the *gradient of the error* at each new trial. Training may therefore be achieved by modifying the classifier f after each trial in the direction *opposite* to the error [WID 62, RUM 88, WIL 92]. If $\tilde{g}(\mathbf{x}, f)$ estimates the error at point $(\mathbf{x}, f)$, the change in f may be written:

$$f \leftarrow f - \eta\tilde{g}(\mathbf{x}, f)$$

where η is a "small" scalar. For example, in the linear case, where $f(\mathbf{x}) = \langle \mathbf{x}, \mathbf{w} \rangle$, the estimator of the gradient of the least squares error is given by $\tilde{g}(\mathbf{x}, \mathbf{w}) = (\langle \mathbf{x}, \mathbf{w} \rangle - y)\, \mathbf{x}$ and the update operation is given by:

$$\mathbf{w} \leftarrow \mathbf{w} - \eta\left(\langle \mathbf{x}, \mathbf{w} \rangle - y\right) \mathbf{x}.$$

where y is the expected response and $(\langle \mathbf{x}, \mathbf{w} \rangle - y)$ is the prediction error.

During usage, BCIs, by definition, do not know the response expected by the subject. The value of the prediction error therefore remains unknown, and classical gradient descent methods are not applicable as such. But we can use a weaker hypothesis stating that we have access to a simple *indication* of the "quality" of the response. For example, in the case of a P300 virtual keyboard, we will be able to tell whether the suggested letter is correct or incorrect, but if the label is wrong, we will not know the correct label. This indication is denoted r; $r = 1$ means that the response is correct, and $r = -1$ means that it is incorrect. Gradient descent methods that are applicable with this setup are based on a probabilistic approach. The output $\{f(\mathbf{x}, i, \mathbf{w})\}_{i \in 1,\ldots,K}$ of the classifier is a probability for each label $1, \ldots, K$, and the response $\tilde{y}$ is the result of a random event with these probability values. The "policy gradient" algorithm [WIL 92] updates the classifier with:

$$\mathbf{w} \leftarrow \mathbf{w} + \eta r \nabla_{\mathbf{w}} \log f(\mathbf{x}, \tilde{y}, \mathbf{w})$$

A numerical study published in [DAU 15] shows that the performance of the policy gradient applied to P300 classifiers is close to that of the logistic gradient with complete information. In BCI applications, the problem at hand is to measure in real time the "value" of the response delivered by the interface. There are various ways of achieving this, such as measuring an "error potential" [FAL 91] that indicates a mismatch between the subject's expectation and the response produced by the algorithm [BUT 06], or using a "backspace" character indicating that the *previous* letter was incorrect [DAU 15], or even language-based statistics indicating that the chosen letter is unlikely given the previous letters.

10.3.3. *Subject-adapted calibration*

Before starting to use a BCI, an initial calibration stage is often necessary to determine certain parameters, or choose between the available options. For example, for SSVEPs, the frequency of the flashes needs to be selected to ensure that it appears in the EEG spectrum with a good signal-to-noise ratio [BRY 13]. For motor imagery, tasks must be found that the user can perform in a reproducible manner, in view of promoting their correct classification.

One traditional approach to the initial calibration stage is to ask the subject to repeat a certain set of tasks a large number of times, corresponding to the different options under consideration. The EEG data thus obtained is analyzed *offline* to determine the best choices for the *online* phase. However, this approach has multiple disadvantages as follows:

– data are acquired uniformly across the different options (the same amount of data for each option), but certain options will be unusable; this time would be better spent gathering data from tasks that can be effectively exploited, which will make it possible to establish better classifiers;

– users receive zero feedback about the interpretation of their cerebral activity; the signals gathered in the phase could potentially be significantly different from the signals in the *online* exploitation phase, since user expectations and tasks can vary strongly between the two phases [SHE 06].

Calibration is expensive in terms of time and attention required from the subject: it is therefore important for calibration to be performed rapidly and efficiently [DOB 09].

10.3.3.1. *Reinforcement learning*

Reinforcement learning is a sequential method that selects the best-performing actions for a chosen objective, given the observed state of the system. For BCIs, this method provides a promising way of exploring the available options, while simultaneously exploiting them online. Reinforcement learning searches for a compromise between exploration and exploitation. Thus, the two disadvantages of traditional precalibration listed above can be corrected: data acquisition is no longer necessarily uniform across the available options (exploration), and data processing is performed online, providing feedback to the user (exploitation). The most simple algorithm of reinforcement learning is the *multiarmed bandit algorithm* [AUE 02]. The algorithm was named after a series of slot machines ("one-armed bandits"), in front of which a player would naturally attempt to maximize profit.

10.3.4. *Optimal tasks*

The multiarmed bandit algorithm has been used to rapidly select from a set of motor imagery tasks the most discriminating task relative to the idle state [FRU 13]. There is a random factor coming from the low signal-to-noise ratio of cerebral signals, which explains why it is necessary to measure multiple trials for each task in order to classify them. Our knowledge of the target function (here, the classification rate that we are attempting to maximize) is imprecise, but we can find bounds that confine it almost surely within a confidence interval. The bounds of the confidence interval are calculated from past observations, and are updated with each new observation. The bandit algorithm performs sequentially: at each stage, it selects from the K possible tasks the task that maximizes an upper bound of the target function. This is illustrated for $K = 2$ in three successive steps in Figure 10.3(top). Over the course of a single experiment, the algorithm chooses a sequence of tasks for the subject to perform, as shown at the bottom of Figure 10.3. Table 10.2 shows the results of five successive optimal task selection experiments with the same subject, each time selecting the best task, with the number of trials (observations) of each class and their final classification rate. We can see that in each of the experiments, feet motion achieved the best classification rate. Tongue motion, which had the lowest classification rate, was performed fewer times than the other tasks.

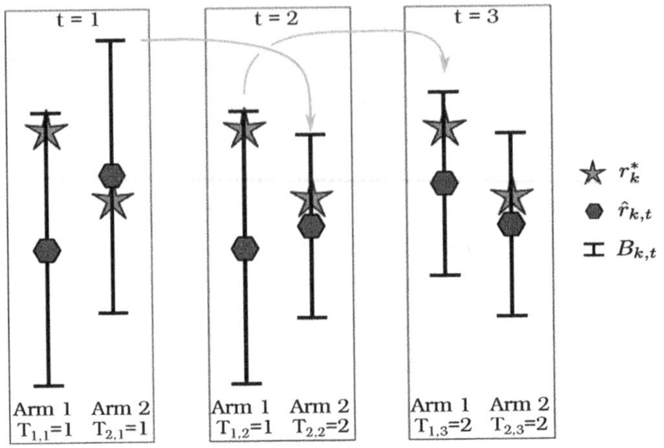

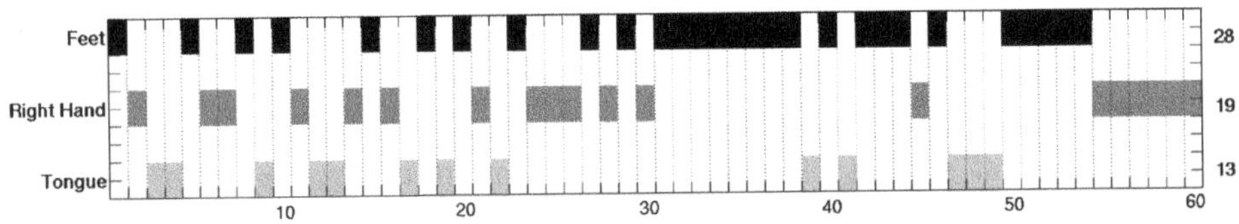

Figure 10.3. *Top: Three stages of the multiarmed bandit algorithm $t \in \{1, 2, 3\}$. Each task $k \in \{1, 2\}$ has a theoretical classification rate r_k^*. At stage t, this rate is estimated by $\hat{r}_{k,t}$ with a confidence interval $B_{k,t}$ whose size decreases as the number of realizations of task k (given by $T_{k,t}$) increases. Bottom: Sequence of selected motor imagery tasks as the bandit method is iterated during an online experiment (last row of Table 10.2); motor imagery of the feet is the most commonly performed task, and the most discriminating. For a color version of this figure, see www.iste.co.uk/clerc/interfaces1.zip*

Right hand		Feet		Tongue		
Number of trials	Classification rate (%)	Number of trials	Classification rate (%)	Number of trials	Classification rate (%)	Selected task
16	80	28	97	16	78	Feet
19	72	30	87	11	40	Feet
16	70	34	86	10	57	Feet
15	64	34	88	11	55	Feet
19	73	28	86	13	76	Feet

Table 10.2. *Results of five successive instances of an experiment selecting the best task using the multiarmed bandit algorithm, performed with the same subject*

10.3.5. *Correspondence between task and command*

The characteristic features of cerebral responses are assigned classes by the classifier, but classes and the commands do not necessarily have a one-to-one correspondence. Although most BCIs fix this correspondence *a priori*, certain studies have suggested allowing the system to learn the relationship in real time. The first study that used reinforcement learning in this way was performed on rats, whose unitary activity in the motor cortex was recorded [DOB 09]. By online optimization of a multilayer perceptron, the rat could access a reward by operating a 3D robot arm toward a predefined target. A more recent study uses partially observable Markov decision processes with SSVEPs [BRY 13]. Users were able to modify the correspondence between three flashing zones and three selectable targets. Unlike the motor study in the rat, this last process was supervised because the sequence of targets to select was known *a priori*.

10.4. Generative model and adaptation

10.4.1. *Bayesian approach*

The Bayesian framework is as simple as it is generic, because it can be applied to any probabilistically formulated inference problem and is based on a single formula, Bayes' law:

$$p(y|\mathbf{x}, M) = \frac{p(\mathbf{x}|y, M)p(y|M)}{p(\mathbf{x}|M)}, \qquad [10.3]$$

where we recall that $\mathbf{x}$ is the vector of observations or relevant features of the signal and y is the hidden variable that we wish to estimate (the command or mental state that defines the interaction). Equation [10.3] explicitly exhibits the dependency on the chosen model M composed of all the hypotheses about the causal process, linking the intent or state y to the observations $\mathbf{x}$.

Defining M is equivalent to defining the distributions $p(\mathbf{x}|y, M)$ (the likelihood of the observations, given the variable y describing the state) and $p(y|M)$ (our *a priori* knowledge about y). Given these distributions, the Bayesian inference step simply applies equation [10.3] to obtain the *a posteriori* distribution $p(y|\mathbf{x}, M)$ and the evidence $p(\mathbf{x}|M)$ in favor of the model M, which is just a (scalar) normalization factor.

Although Bayes' law itself is simple, its application can prove difficult if the model M is nonlinear, or if the distributions that define it cannot be solved analytically. In these cases, it is necessary to resort to approximation techniques; these techniques can provide numerical solutions that are precise but expensive to calculate (this is the case for sampling methods [ROB 04]), or analytical solutions that are very efficient computationally but potentially less precise (this is the case for variational methods [FRI 07]).

But, more generally, the states y are not the only hidden variables. The model M usually includes parameters θ that must be estimated, specific to each subject or patient. The equation to be solved then becomes:

$$p(y, \theta|\mathbf{x}, M) = \frac{p(\mathbf{x}|y, \theta, M)p(y, \theta|M)}{p(\mathbf{x}|M)} \qquad [10.4]$$

The purpose of the θ variable is to make certain aspects of the model M explicit, such as the number and the nature of the unknown parameters. But M contains other information as well, such as the nature of the relationship between the hidden parameters and the observed data. For example, when reconstructing the sources for EEG or MEG, models known as distributed models describe cortical activity using a large number of current dipoles with fixed position and orientation. Thus, in addition to the θ parameters describing the amplitude of these dipoles, the model M specifies other hypotheses such as their positions and orientations, which are defined using a reference brain, or from the MR images of the subject [MAT 07].

Besides the fact that the state variable is the primary variable of interest, as it determines the conditions of the interaction, its main difference with the other unknown parameters θ is that it varies over time. Usually, the θ parameters are considered known, once estimated. They can be estimated in advance, e.g. during calibration. If we consider the example of the P300 speller, θ is the vector of parameters corresponding to each class (target and non-target) for a given user, and y is the label assigned to the observation x. But equation [10.4] shows that the θ parameters may also be estimated,jointly with y. This highlights the possibility that the Bayesian approach could be used to develop approaches that would be capable of adapting the model parameters to each user.

Once Bayesian inference has been performed, the *a posteriori* probability distribution $p(y, \theta|\mathrm{x}, M)$ allows the hidden states and the parameters of the model to be inferred. This distribution provides an estimate of the correct command for the BCI. Note that this *a posteriori* estimate, expressed as a probability distribution, gives not just an estimate of the (average) value of these variables, but also an estimate of their variance, which is a measure of the confidence in the estimate. The variance can be used, for example by setting up an adaptive decision-making strategy for optimal stopping [MAT 15] (see section 10.4.2).

The evidence $p(\mathbf{x}|M)$ also allows us to make inferences about the models, and potentially to compare different, alternative models [MAT 07] (see section 10.4.3).

BCIs are essentially interested in estimating states and sometimes parameters, so they focus primarily on the first type of inference. With this objective in mind, equations [10.3] and [10.4] show that Bayes' law is just a rule for updating the state of knowledge in the presence of new observations. Since BCIs work in real time and gather observations sequentially, Bayesian reasoning, applied to each new observation, establishes a set of rules for learning and endows the machine with the capacity to adapt. After explicitly introducing a notion of time to index the sequential events or trials t, Bayesian learning may be written as follows:

$$p(y|\mathbf{x}_t, M) \propto p(\mathbf{x}_t|y, M)p(y|M) \qquad [10.5]$$

where the *a priori* distribution of the state y is given by the *a posteriori* distribution learned from earlier observations

$$p(y|M) = p(y|\mathbf{x}_1...\mathbf{x}_{t-1}, M) \qquad [10.6]$$

Many of the existing methods of feature extraction or classification can be reformulated within the Bayesian framework. Conversely, the Bayesian framework, and in particular the explicit formulation of a generative model for the data, allows for *a priori* information to be integrated into a specific context of experimental interaction, thus making it possible to optimize and explore the possibility of identifying parameters directly online. For example, with the P300 speller, the inference and dynamic decision process may be constrained using a language model that provides *a priori* information about letters and the most probable letter sequences [MAI 15].

10.4.2. *Sequential decision*

It is desirable for the BCI to only make a decision when the confidence in its correctness is sufficiently high. To achieve this, strategies for evidence accumulation have been suggested [VER 12, DAU 14], which classify the user's cerebral activity over a variable time duration, and output a decision when the evidence has exceeded a chosen threshold.

This problem may also be expressed in probabilistic terms. Consider the problem of finding the number of measurements necessary to find the (unique) label $y \in 1, ..., K$ shared by a series of observations $\mathbf{x}_1, ..., \mathbf{x}_t, ...$. If $P(y = k|\mathbf{x}_1, ..., \mathbf{x}_{t-1})$ is known, the probability assigned to the new measurement $\boldsymbol{x}_t$ is given by:

$$P(y = k|\mathbf{x}_1, ..., \mathbf{x}_t) = \frac{P(\mathbf{x}_t|y = k)P(y = k|\mathbf{x}_1, ..., \mathbf{x}_{t-1})}{\sum_{i=1}^{K} P(\mathbf{x}_t|y = i)P(y = i|\mathbf{x}_1, ..., \mathbf{x}_{t-1})} \quad [10.7]$$

Starting from a known initial distribution $P(y = k), k = 1, ..., K$, we can sequentially refine the assigned probability of each label with each new measurement. The sequential decision problem [WAL 47] consists of choosing the optimal moment to terminate the measurement process, given a fixed decision threshold s. The "optimal stopping" algorithm 10.1 allows us to decide after each measurement whether to continue or terminate the measurement process.

Algorithm 10.1. Optimal stopping (s)

```
1: P_1 ← P(y = 1), ..., P_K ← P(y = K)
2: while ∄i : P_i ≥ s do
3:     Perform a measurement x
4:     Update P_1, ..., P_K
5: end while
```

In the example of the virtual keyboard "P300", we denote $\mathbf{x}_1, ..., \mathbf{x}_T$ the sequence of observations obtained from T flashes. The label $y \in 1..K$ corresponds to the target letter (where K is the total number of listed of proposed symbols). We denote $P^+(\mathbf{x}_t)$ the probability that observation $\mathbf{x}_t$ is a P300. If we know the list S_t of letters flashed on the screen at each instant t, we can construct the *a posteriori* instantaneous probability of each symbol [DAU 14], which is:

$$P(y = k|\mathbf{x}_t) = \frac{1}{n}\mathbb{I}_{k \in S_t} P^+(\mathbf{x}_t) + \frac{1}{K-n}\mathbb{I}_{k \notin S_t}(1 - P^+(\mathbf{x}_t)) \qquad [10.8]$$

where n is the number of highlighted symbols for each flash.

The value of $P(y = k|\mathbf{x}_1, ..., \mathbf{x}_t)$ may then be deduced from equation [10.7] using a uniformity hypothesis, namely $\exists C : \forall k, P(y = k|\mathbf{x}_t) = CP(\mathbf{x}_t|y = k)$. An optimal stopping strategy, as illustrated in Figure 10.4, may then be implemented.

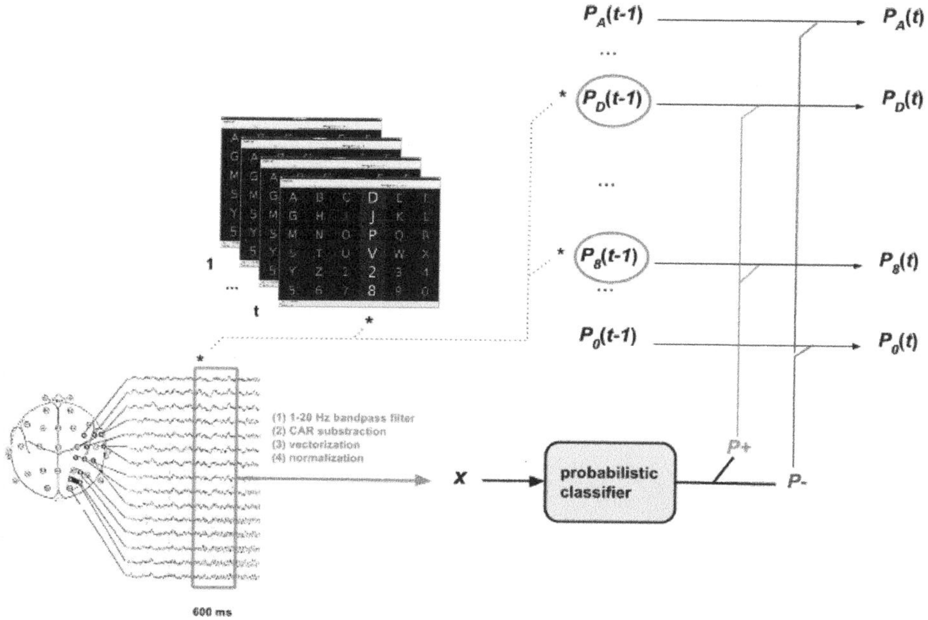

Figure 10.4. *Evidence accumulation with the P300 speller. Each flash is assigned a probability P^+ of eliciting a P300, which allows the vector of probabilities associated with the various different symbols of the virtual keyboard to be updated (see text). For a color version of this figure, see www.iste.co.uk/clerc/interfaces1.zip*

10.4.3. *Online optimization of stimulations*

Sequential hypothesis testing does not just involve determining the optimal moment to make a decision and stop gathering information. The nature of the test must also be optimized, that is to say the nature of the question being asked, or, in the case of reactive BCIs such as the P300 speller, the identity of the flashing items. To see this, we can simply make explicit the role of the stimulation u_t at an instant t when the observed data are generated. With the Bayesian formalism, this leads us to rewrite equation [10.5] in the following form:

$$p(y|\mathbf{x}_t, M_k, u_t) = \frac{p(\mathbf{x}_t|y, M_k, u_t)p(y|M_k)}{p(\mathbf{x}_t|M_k, u_t)} \quad [10.9]$$

where M_k is the kth alternative hypothesis.

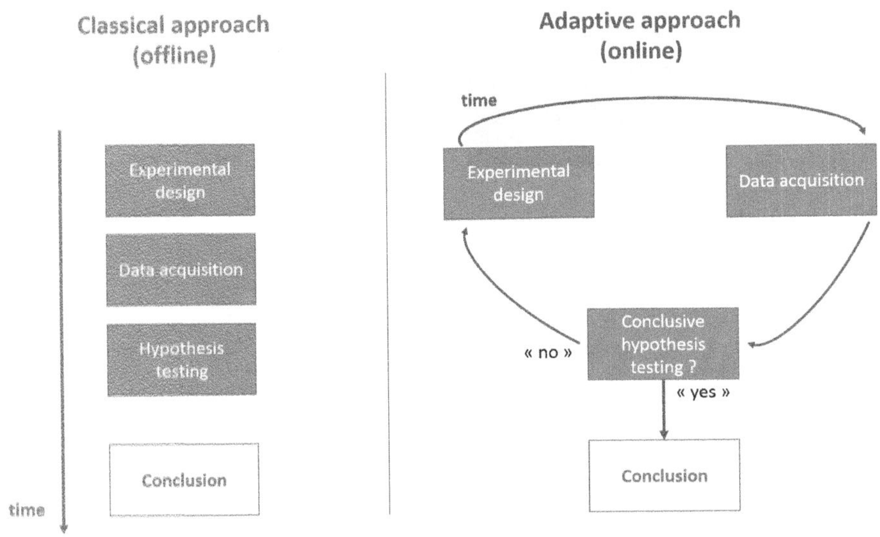

Figure 10.5. *General principles of the traditional experimental design (left panel) and the adaptive experimental design (right panel)*

As explained when it was introduced in section 10.4.1, the Bayesian approach allows us to make inferences about the models, by formally comparing several alternative hypotheses. *In fine*, the chosen model will be the model that has an *a posteriori* probability above some chosen threshold (typically 0.95). At time t, for the model M_k, this probability is given by:

$$p(M_k|\mathbf{x}_1...\mathbf{x}_t, u_1...u_t) = \frac{1}{1 + \sum_{M_i \neq M_k} BF_{ik}} \qquad [10.10]$$

where BF_{ik} is the Bayes factor between models M_i and M_k:

$$BF_{ik} = \frac{p(\mathbf{x}_1...\mathbf{x}_t|M_i, u_1...u_t)}{p(\mathbf{x}_1...\mathbf{x}_t|M_k, u_1...u_t)} \qquad [10.11]$$

Given this criterion for comparing hypotheses, which can also be used as an optimal stopping criterion, the choice of the next optimal stimulation u_{t+1} must satisfy the condition that the risk of error in the model selection is minimized [DAU 11]. Figure 10.6 shows the performance increase in realistic Monte Carlo simulations achieved by online adaptation of the stimulations, compared to the classical approach in which the objects are flashed in a random order.

10.4.3.1. *Adaptive experiments in cognitive neuroscience*

Optimizing stimulations, or in other words optimizing the experimental design, is a central challenge in cognitive neuroscience, where the objective of an experiment is usually to compare alternative hypotheses on brain function, whether healthy or pathological. These hypotheses effectively model the relationship between mental processes (learning, decision making, etc.) and neurophysiological or behavioral responses [STE 15]. In computational neuroscience, these models have become more realistic but also more complex; it has become particularly difficult to preoptimize the experimental design so that they may be compared, for a given subject [DAU 11].

Thanks to BCI developments, it is now possible to analyze cerebral activity in real time. Optimization is achieved by adapting directly online the successive stimulations used during experiments in cognitive neuroscience and neuroimaging [SAN 14].

In practice, implementing this adaptive approach involves fitting the data with each of the considered models for each new observation, evaluating the evidence in favor of each model and calculating the discriminative power for deciding between the models of each of the possible values of the upcoming stimulation u. This approach represents a departure from traditional protocols, which fix the experimental design in advance. It provides an alternative strategy, optimized for hypothesis testing, and produces an optimal and specific paradigm for each new subject or patient. This is illustrated in Figure 10.5.

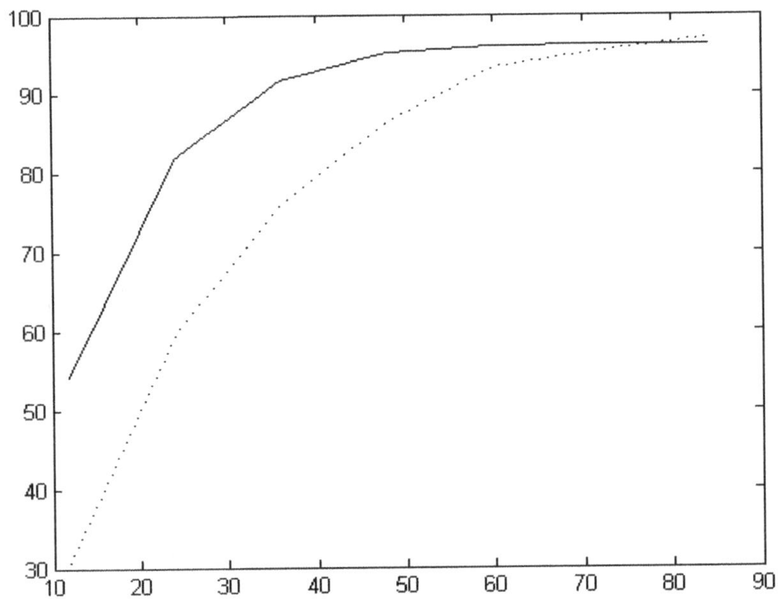

Figure 10.6. *Percentage of correctly spelled letters, for simulated P300 speller data (N = 2,000) as a function of the number of stimulations in the case where the stimulations are ordered pseudo-randomly such that each letter appears an equal number of times (dotted line), and in the case where the choice of stimulation is made online to optimize the detection of the target letter (solid line)*

However, the validity of such an approach remains to be empirically demonstrated. One major obstacle shared by all BCIs is the problem of removing possible artifacts in real time, which traditionally would be performed in fine detail offline. Another example of a possible application for this adaptive approach is the field of diagnostics and prognostics for patients with perturbed states of consciousness, as mentioned in Chapter 1 of Volume 2 [CLE 16].

10.5. Conclusions

We have presented a wide range of methods of adaptive learning, grouped into two families: methods that perform statistical decoding and methods based on a generative model. These methods have not yet found widespread application with BCIs, and the future promises development and progress; the necessity of being able to adapt to the variability of the recorded signals regardless of the origin of this variability can be clearly recognized today. Note that these methods are not necessarily mutually exclusive. Indeed, future developments will most likely draw from the strengths of each of them. Whatever their final form may be, these methods will likely be built with ever-improving mathematical tools suitable for online deployment and will rely on future improvements in our understanding of neurophysiology and cognitive neuroscience, allowing us to create more efficient classification functions or generative models with appropriate adaptive capabilities.

10.6. Bibliography

[AMI 13] AMIEZ C., NEVEU R., WARROT D. *et al.*, "The location of feedback-related activity in the midcingulate cortex is predicted by local morphology", *Journal of Neuroscience*, vol. 33, no. 5, pp. 2217–2228, 2013.

[AUE 02] AUER P., CESA-BIANCHI N., FISCHER P., "Finite time analysis of the multiarmed bandit problem", *Machine Learning*, vol. 47, nos.2-3, pp. 235–256, 2002.

[BER 07] BERKA C., LEVENDOWSKI D.J., LUMICAO M.N. *et al.*, "EEG correlates of task engagement and mental workload in vigilance, learning, and memory tasks", *Aviation, Space, and Environmental Med*icine, vol. 78, no. 5, pp. B231–B244, 2007.

[BOM 15] BOMPAS A., SUMNER P., MUTHUMUMARASWAMY S.D. *et al.*, "The contribution of pre-stimulus neural oscillatory activity to spontaneous response time variability", *NeuroImage*, vol. 107, pp. 34–45, 2015.

[BRY 13] BRYAN M.J., MARTIN S.A., CHEUNG W. *et al.*, "Probabilistic co-adaptive brain–computer interfacing", *Journal of Neural Engineering*, vol. 10, no. 6, 2013.

[BUT 06] BUTTFIELD A., FERREZ P., DEL R.M.J., "Towards a robust BCI: error potentials and online learning", *IEEE Transactions on Neural Systems and Rehabilitation Engineering*, vol. 14, no. 2, pp. 164–168, 2006.

[CLE 16] CLERC M., BOUGRAIN L. LOTTE F., *Brain–Computer Interface 2*, ISTE, London and John Wiley & Sons, New York, 2016.

[DAL 97] DALEBOUT S.D., ROBEY R.R., "Comparison of the intersubject and intrasubject variability of exogenous and endogenous auditory evoked potentials", *Journal of American Academy of Audiol*ogy, vol. 8, no. 5, pp. 342–354, 1997.

[DAU 11] DAUNIZEAU J., PREUSCHOFF K., FRISTON K. *et al.*, "Optimizing experimental design for comparing models of brain function", *PLoS Computational Biology*, vol. 7, no. 11, p. e1002280, 2011.

[DAU 14] DAUCÉ E., THOMAS E., "Evidence build-up facilitates on-line adaptivity in dynamic environments: example of the BCI P300-speller", in VERLEYSEN M., (ed.), *Proceedings of the 22nd European Symposium on Artificial Neural Networks, Computational Intelligence and Machine Learning (ESANN)*, Bruges, Belgium, pp. 377–382, April 23–25, 2014.

[DAU 15] DAUCÉ E., PROIX T., RALAIVOLA L., "Reward-based online learning in non-stationary environments: adapting a P300-speller with a "Backspace" key", *Proceedings of the International Joint Conference on Neural Networks (IJCNN)*, Killarney, Ireland, July 12–17, 2015.

[DEM 77] DEMPSTER A.P., LAIRD N.M., RUBIN D.B., "Maximum likelihood from incomplete data via the EM algorithm", *Journal of the Royal Statistical Society, Series B (Methodological)*, vol. 39, no. 1, pp. 1–38, 1977.

[DIG 09] DIGIOVANNA J., MAHMOUDI B., FORTES J. *et al.*, "Co-Adaptive Brain-Machine Interface via Reinforcement Learning", *IEEE Transactions on Biomedical Engine*ering, vol. 56, no. 1, pp. 54–64, 2009.

[DOB 09] DOBREA M.-C., DOBREA D., "The selection of proper discriminative cognitive tasks – a necessary prerequisite in high-quality BCI applications", *2nd International Symposium on Applied Sciences in Biomedical and Communication Technologies, (ISABEL'09)*, pp. 1–6, 2009.

[DRE 11] DREWES J., VANRULLEN R., "This is the rhythm of your eyes: the phase of ongoing electroencephalogram oscillations modulates saccadic reaction time", *Journal of Neuroscience*, vol. 31, no. 12, pp. 4698–4708, 2011.

[FAL 91] FALKENSTEIN M., HOHNSBEIN J., HOORMANN J. *et al.*, "Effects of crossmodal divided attention on late ERP components. II. Error processing in choice reaction tasks", *Electroencephalography and Clinical Neurophysiology*, vol. 78, no. 6, pp. 447–455, 1991.

[FRI 02] FRISTON K., GLASER D., HENSON R., KIEBEL S. *et al.*, "Classical and bayesian inference in neuroimaging: applications", *NeuroImage*, vol. 16, no. 2, pp. 484–512, 2002.

[FRI 07] FRISTON K., MATTOUT J., TRUJILLO-BARRETO N. *et al.*, "Variational free energy and the Laplace approximation", *NeuroImage*, vol. 34, no. 1, pp. 220–234, 2007.

[FRU 13] FRUITET J., CARPENTIER A., MUNOS R. *et al.*, "Automatic motor task selection via a bandit algorithm for a brain-controlled button", *Journal of Neural Engineering*, vol. 10, no. 1, 2013.

[GRA 91] GRAFTON S., WOODS R., MAZZIOTTA J. *et al.*, "Somatotopic mapping of the primary motor cortex in humans: activation studies with cerebral blood flow and positron emission tomography", *Journal of Neurophysiol*ogy, vol. 66, no. 3, pp. 735–743, 1991.

[HAE 14] HAEGENS S., COUSIJN H., WALLIS G. *et al.*, "Inter- and intra-individual variability in alpha peak frequency", *NeuroImage*, vol. 92, pp. 46–55, 2014.

[KIN 12a] KINDERMANS P.-J., VERSCHORE H., VERSTRAETEN D. *et al.*, "A P300 BCI for the masses: prior information enables instant unsupervised spelling", *Advances in Neural Information Processing Systems*, pp. 710–718, 2012.

[KIN 12b] KINDERMANS P.-J., VERSTRAETEN D., SCHRAUWEN B., "A Bayesian model for exploiting application constraints to enable unsupervised training of a P300-based BCI", *PLoS One*, vol. 7, no. 4, p. e33758, 2012.

[KIV 04] KIVINEN J., SMOLA A.J., WILLIAMSON R.C., "Online learning with kernels", *IEEE Transactions on Signal Processing,* vol. 52, no. 8, pp. 2165–2176, 2004.

[KOH 07] KOHN A., "Visual adaptation: physiology, mechanisms, and functional benefits", *Journal of Neurophysiol*ogy, vol. 97, no. 5, pp. 3155–3164, 2007.

[LI 06] LI Y., GUAN C., "An extended EM algorithm for joint feature extraction and classification in brain–computer interfaces", *Neural Computation*, vol. 18, no. 11, pp. 2730–2761, 2006.

[LI 10] LI Y., KAMBARA H., KOIKE Y. *et al.*, "Application of covariate shift adaptation techniques in brain–computer interfaces", *IEEE Transactions on Biomedical Engineering,* vol. 57, no. 6, pp. 1318–1324, 2010.

[MAI 15] MAINSAH B.O., MORTON K.D., COLLINS L.M. *et al.*, "Moving away from error-related potentials to achieve spelling correction in P300 spellers", *IEEE Transactions on Neural Systems and Rehabilitation Engineering*, vol. 23, no. 5, pp. 737–743, 2015.

[MAT 07] MATTOUT J., HENSON R.N., FRISTON K.J., "Canonical source reconstruction for MEG", *Computational Intelligence and Neuroscience*, vol. 2007, pp. 1–10, 2007.

[MAT 15] MATTOUT J., PERRIN M., BERTRAND O. *et al.*, "Improving BCI performance through co-adaptation: Applications to the P300-speller", *Annals of Physical and Rehabilitation Medicine*, vol. 58, no. 1, pp. 23–28, 2015.

[MOR 14] MORLET D., FISCHER C., "MMN and novelty P3 in coma and other altered states of consciousness: a review", *Brain Topography*, vol. 27, no. 4, pp. 467–479, 2014.

[ROB 04] ROBERT C.P., CASELLA G., *Monte Carlo Statistical Methods*, Springer, New York, NY, 2004.

[RUM 88] RUMELHART D.E., HINTON G.E., WILLIAMS R.J., "Learning representations by back-propagating errors", *Cognitive Modeling*, vol. 5, pp. 3, 1988.

[SAN 14] SANCHEZ G., DAUNIZEAU J., MABY E. *et al.*, "Toward a new application of real-time electrophysiology: online optimization of cognitive neurosciences hypothesis testing", *Brain Sciences*, vol. 4, no. 1, pp. 49–72, 2014.

[SHE 06] SHENOY P., KRAULEDAT M., BLANKERTZ B., RAO R. *et al.*, "Towards adaptive classification for BCI", *Journal of Neural Engineering*, vol. 3, no. R13, 2006.

[STE 15] STEPHAN K., IGLESIAS S., HEINZLE J. *et al.*, "Translational perspectives for computational neuroimaging", *Neuron*, vol. 87, no. 4, pp. 716–732, 2015.

[SUG 07] SUGIYAMA M., KRAULEDAT M., MÜLLER K., "Covariate shift adaptation by importance weighted cross validation", *Journal of Machine Learning Research*, vol. 8, pp. 985–1005, 2007.

[VER 12] VERSCHORE H., KINDERMANS P.-J., VERSTRAETEN D. *et al.*, "Dynamic stopping improves the speed and accuracy of a P300 speller", *Artificial Neural Networks and Machine Learning–ICANN'12*, pp. 661–668, Springer, 2012.

[VID 11] VIDAURRE C., SANNELLI C., MÜLLER K.-R. *et al.*, "Co-adaptive calibration to improve BCI efficiency", *Journal of Neural Engineering*, vol. 8, no. 2, 2011.

[WAL 47] WALD A., *Sequential Analysis*, Dover Phoenix Editions, Mineola, NY, 1947.

[WEI 06] WEISSMAN D.H., ROBERTS K.C., VISSCHER K.M. *et al.*, "The neural bases of momentary lapses in attention", *Nature Neuroscience*, vol. 9, no. 7, pp. 971–978, 2006.

[WID 62] WIDROW B., "Generalization and information storage in network of adaline'neurons"', in YOVITZ M.C. , JACOBI G.T., GOLDSTEIN G. (eds), *Self-Organizing Systems-1962*, Spartan Books, pp. 435–462, 1962.

[WIL 92] WILLIAMS R.J., "Simple statistical gradient-following algorithms for connectionist reinforcement learning", *Machine Learning*, vol. 8, nos. 3–4, pp. 229–256, 1992.

11

Human Learning for Brain–Computer Interfaces

11.1. Introduction

BCIs are defined by Wolpaw [WOL 02] as tools of communication and control that allow users to interact with their environment by means of their cerebral activity alone. This definition highlights one fundamental aspect of BCIs, the interaction between two components: the user's brain and the computer. The challenge is to make sure that these two components (brain and computer) "understand each other", and adapt to each other so that the system performance (often measured using the accuracy rate) is optimal.

Thus, the working architecture of a BCI [WOL 02] contains a loop with two major stages, after the user sends a command via cerebral activity (which we shall denote stage 0). During stage I, the computer attempts to *understand* the command sent by the user, generally by extracting relevant information followed by classification. Next, during stage II, it is the user's turn to attempt to *understand* the meaning of the feedback generated by the computer, which indicates how the computer understood the command that it received. To see how this loop works, consider the case of a standard BCI protocol in motor imagery [PFU 01]. In this protocol, users can perform two motor imagery tasks, "imagine moving the left hand" and "imagine moving the right hand",

Chapter written by Camille JEUNET, Fabien LOTTE and Bernard N'KAOUA.

which are associated with two distinct commands. To provide guidance to the user, the system also produces feedback, often in the form of a bar indicating the task recognized by the system. The direction of the bar depends on the task recognized by the system (e.g. the bar points left if the task "imagine moving the left hand" is recognized). The size of the bar also depends on the value of the classifier output (i.e. higher values indicate that the classifier is more confident in the task recognition, and so the bar will be larger) (see Figure 11.1, left-hand side). In this example, stage I of the loop is the computer's recognition of the motor imagery task performed by the user (is the user imagining moving his left hand, or his right hand?). Then, in stage II, the user now has to understand the feedback generated by the system (what does this bar mean? Did the system correctly recognize the task that I performed? If so, how confident was it?). Unfortunately, it appears that most current systems do not properly establish this mutual understanding, which might explain why users perform poorly when attempting to control the BCI, as well as the non-negligible fraction (between 15% and 30%) of users that find themselves completely incapable of controlling these systems [ALL 10].

How can we facilitate this understanding? Over the last several years, there have been many studies on stage I of the loop: how the computer should understand the task performed by the user. Signal processing algorithms and techniques of machine learning have been developed to achieve this. But two fundamental factors for improving BCI performance have not yet been sufficiently explored:

– *Stage 0, the quality of the signals generated by the user*: for the classification algorithms to be effective (i.e. in order that they can be capable of recognizing motor imagery tasks by extracting specific features from the cerebral signal), the user must be able to generate a *stable* cerebral signal each time that he/she performs the same task, and *distinct* cerebral signals when the tasks are different. These two elements are non-trivial skills, and require a learning process that is both specific and adapted. This is rarely taken into account in BCI training protocols [NEU 10].

– *Stage II, user comprehension of the feedback produced by the system*: The standard BCI protocols often provide the user with feedback in the form of a graphical representation of the classifier output (e.g. the bar described above). Although this is informative (and more importantly allows evaluation/correction), this feedback does not explain to the user why a certain task was or was not recognized, and even less what the user must do in order to improve performance. In a recent review [LOT 13], Lotte *et al.* show that to be effective feedback must provide an explanation (rather than just the possibility of correction), be multimodal (and not just visual), and finally be clear and explicit (which is not the case with classifier output for non-experts).

These different ideas highlight a point that might allow user performance to be improved: facilitating the acquisition of skills by providing adapted learning protocols. As we will see in this chapter, establishing a learning protocol requires various different elements to be taken into consideration: the instructions/indications given to the user, the learning environment, the practice tasks given to the learner and the feedback provided after performing the various different tasks.

In section 11.2, we will explore the limitations of the standard protocols widely used by the BCI community. Next, we will analyze the learning protocols that have been suggested for BCIs. We will focus on protocols developed for teaching users how to use BCIs based on mental imagery (MI), also known as spontaneous BCIs. Indeed, this is the category of BCI for which the learning process is the most important. Finally, in section 11.4, we will present possible avenues for improving learning protocols, in particular based on an "anthropocentric" perspective. Before we begin, however, let us describe two *historical* approaches that were used with BCIs, on which most of the current learning protocols are based. One protocol was suggested by researchers in Graz [PFU 01] based on techniques of *machine learning*, and the other was suggested by the researchers at the Wadsworth center [WOL 00] based on an *operant conditioning* approach.

11.2. Illustration: two historical BCI protocols

Principle of the Graz protocol [PFU 01]: This approach is organized into two stages: I: training the system; II: teaching the user. In stage I, the user is instructed to successively perform a certain series of MI tasks (for example imagining movements of the left and right hands). Using the recordings of cerebral activity generated as these various MI tasks are performed, the system attempts to extract characteristic patterns of each of the mental tasks from the signal (see Chapter 7). These extracted features are used to train a *classifier* whose goal is to determine the *class* to which the signals belong (i.e. imagining a movement of the left hand or the right hand) (see Chapter 12). This classifier is then typically adjusted over the course of the learning session so that variations in the disposition of the apparatus or in the user conditions between sessions are taken into account. When this stage is complete, stage II involves training the user. The user is instructed to perform the MI tasks, but this time feedback (based on the learning performed by the system in stage I) is provided to inform him or her of the MI task recognized by the system and the corresponding confidence level of the classifier. The user's goal is to develop effective *strategies* that will allow the system to easily recognize the MI tasks that the user is performing.

Definitely, this learning protocol is generally organized over multiple sessions, each of which is composed of sequences (often called *runs*) lasting approximately 7 min. Each session generally has four to six sequences to avoid fatigue, which is often observed after the sixth sequence. Finally, the sequences themselves are divided into trials. One sequence contains 10–20 trials per class (i.e. per MI task) depending on the number of classes. A trial typically lasts for 8 s, during which time a cross appears on the screen followed by a sound to attract the user's attention, further followed by an arrow symbolizing the instruction (e.g. an arrow pointing to the left corresponds to the instruction "imagine moving the left hand") and then visual feedback shown as a bar indicating the recognized task and the corresponding confidence interval of the classifier (e.g. a blue bar pointing to the left means that the system recognized the task of imagining a movement of the left hand; the length of the bar indicates the confidence of the classifier in the recognition of the MI task). The detailed chronology of a trial is shown in Figure 11.1(left).

Principle of the Wadworth center protocol for one-dimensional (1D) cursor control: the BCI system suggested by the Wadsworth center team is based on controlling the sensorimotor rhythms μ and β after a learning process based on operant conditioning [WOL 00]. The initial version of this BCI system, which has now become standard, featured a cursor (or ball) on the screen moving continuously from the left to the right of the screen, at constant speed. The user can control the vertical position of the cursor by modulating the amplitude of his or her sensorimotor rhythms. On the right-hand side of the screen, several targets (generally between two and four, represented by rectangles) are shown, aligned vertically, one by one. The user must adjust the vertical position of the cursor using the BCI so that the cursor hits the indicated target when it reaches the right-hand edge of the screen (see Figure 11.1(right)). This kind of BCI, based on operant conditioning, does not impose any specific mental task on the user, unlike the BCI approach from Graz, nor does it make use of machine learning. Users must find the strategy that allows them to effectively modulate their cerebral rhythms to move the cursor across the screen, on their own. Typically, users utilize motor imagery tasks at the beginning of the learning process, but with practice they report that they use these motor imagery tasks less and less [WOL 00]. Learning to control the BCI takes time, generally several days, weeks or even months of practice. This principle has nonetheless enabled certain users to master controlling a cursor with this BCI in 1D [WOL 00], two dimensions (2D) [WOL 00], and more recently even three-dimensions (3D) [MCF 10]. Notably, this approach was used in the renowned study by Birbaumer *et al.* [BIR 99] published in *Nature* in 1999.

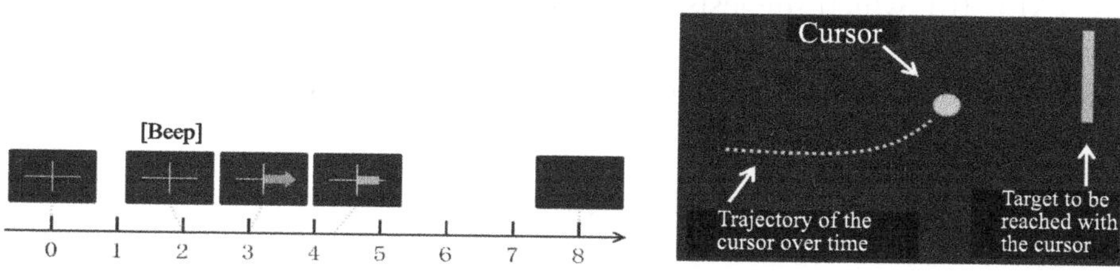

Figure 11.1. *Left: Chronology of a trial: at the beginning of the trial, a cross appears at the center of the screen; after 2 s, a sound is played to indicate that the instruction is imminent; at 3 s, an arrow appears for 1.25 s: the direction of the arrow indicates the MI task that should be performed; at 4.25 s, the feedback is shown for 4 s, and is generally updated 16 times per second depending on classifier output. Right: schematic illustration of a learning trial with the Wadsworth center protocol [WOL 00]*

11.3. Limitations of standard protocols used for BCIs

It has been shown that the standard protocols do not follow the suggested guidelines for learning processes, in particular those suggested by experts in the psychology of learning [LOT 13]. But can we be sure that this has a tangible impact on the user performance when controlling a BCI? Indeed, many other issues have been raised with BCIs, including material problems (e.g. EEG: low signal/noise ratio, variable impedance over time), software problems (e.g. imperfect classification algorithms), and even neurophysiological problems (e.g. non-stationarity of signals). So how can we assess the real impact of learning procedures on BCI performance? In an attempt to answer this question, we used a standard BCI learning protocol [PFU 01] to teach users how to perform a series of simple motor tasks [JEU 14] in a setting without BCIs, and therefore free of all of their associated problems. Over the course of the experiment, users learned how to draw circles and triangles on a digital drawing tablet. Using the same principle as BCI learning, they were instructed to identify the correct strategy (size of the drawing and speed of execution) that allowed the system to recognize the task that they were performing. The results of this study show that, even if the majority does learn and improve over the course of the sequences, 15% of participants obtain accuracy rates of approximately 50% (which is due to randomness, and therefore means that they did not manage to learn to perform these simple motor tasks). This percentage is close to the rate of "illiteracy" in BCIs [ALL 10], which suggests that these protocols are not perfectly adapted to allow users to acquire skills. This result proves the benefit of improving learning protocols with the objective of optimizing BCI performance.

11.4. State-of-the-art in BCI learning protocols

This section offers a review of the literature on existing BCI learning protocols with the objective of establishing guidelines that will be useful for the development of future BCI learning protocols.

11.4.1. *Instructions*

Very few studies have examined the instructions given to users learning to control a BCI. Yet this is a central element of the learning process, since these instructions help users to understand their tasks. Often, these instructions consist only of a single directive indicating that the goal of the exercise is to move the cursor/bar in the right direction. However, as pointed out by Lotte *et al.* [LOT 13], the ultimate objective of the learning protocol is not to move the bar, but to help the user to learn to generate a stable, specific signal for each of the MI tasks that he or she performs. It seems therefore that the learning objective should be made more explicit. One study shows that prompting users to attempt kinesthetic imagination of movements (i.e. to imagine performing the motion, feeling the same sensations, without actually moving anything) rather than simply visual imagination improves the performance [NEU 05]. On the other hand, another study shows that the users that obtained the best performances were those who were not given any specific strategy at the beginning of the learning process [KOB 13]. The authors reason that the success of the learning process depends on subconscious learning mechanisms, and that users who attempt to follow a strategy overload their cognitive resources (which does not result in a positive performance improvement).

11.4.2. *Training tasks*

Although most BCI learning protocols only used one single type of task, which is repeated identically multiple times, a few studies have explored a more varied selection of different tasks. In particular, McFarland *et al.* successfully implemented a progressive sequence of tasks; with operant conditioning, they taught users to first of all control a 1D cursor separately in three different dimensions, then in 2D (for each pair of dimensions), and finally in 3D [MCF 10]. Vidaurre *et al.* experimented with adaptive training tasks by giving subjects a BCI that was initially generic in nature (i.e. independent of the subject, calibrated with the data from multiple other subjects), then progressively more and more adapted to the new user (by adapting the choice of sensors and classifier to this user) [VID 10]. This progressive and co-adaptive approach (the user adapts to the machine and the machine adapts to the user) allowed users that were "illiterate" at first to eventually succeed in controlling the BCI. In a less formal and systematic setting, Neuper *et al.* also explored the idea of allowing the user to learn freely and asynchronously from time to time, with positive results [NEU 03]. Even though this approach has not been compared with the traditional approach (synchronous only), this nonetheless suggests that organizing free-access and asynchronous sessions can be beneficial to BCI learning processes. Finally, Eskandari *et al.* taught their users to meditate before using a BCI, and demonstrated that this had a positive impact on the performance [ESK 08].

11.4.3. *Feedback*

In the standard learning protocols [PFU 01], feedback is given in the form of a bar or a cursor shown on screen, whose direction depends on the task recognized by the classifier and whose size is proportional to the confidence of the classifier in the recognized task. Some studies have suggested other variants for displaying feedback. First of all, Kübler *et al.* [KÜB 01] developed a process that displays a smiley face after each successful trial. In their own study, Leeb *et al.* [LEE 07] replaced the cursor with a gray smiley face that moves toward the left or the right depending on the classifier output. After each trial, the smiley face becomes green and happy if the trial is successful, sad and red if not. This study showed that increasing the motivation levels of users is linked to improved performance. However, neither of these studies offered a formal comparison with the standard feedback process, which makes it impossible to affirm that these kinds of feedback are more effective.

Although the feedback described above (all of which was visual in nature) is simple to implement and intuitive, its effectiveness is not optimal for BCIs. Indeed, it is a recognized fact that in situations of real-life interactions, visual channels are often overloaded [LEE 13], which prompted certain researchers to consider providing feedback via the other senses. Accordingly, several experiments were performed to evaluate the effectiveness of auditory feedback. In the same way as the standard visual feedback, the auditory feedback provided usually represents the classifier output: instead of varying the size of a bar, the classifier output is represented by variations in the frequency of the sound [GAR 12], or its volume [MCC 14], or tone [HIN 04, NIJ 08]. For example, with their *auditory BCI*, Nijboer *et al.* [NIJ 08] used the sounds of two different instruments to indicate the recognition of each of the MI tasks. Although its utility has been proven for patients suffering from locked-in syndrome [SMI 05], because this syndrome is often linked with visual deficiencies and a loss of sensitivity, the performance achieved with auditory feedback has generally been significantly inferior to the performance achieved with visual feedback. One suggested explanation is that it is less intuitive, and thus is longer and more difficult to learn. Also, for real-life applications in open environments (e.g. navigating a wheelchair), the auditory channel is very frequently used and must remain available (much like the visual channel). These factors suggest that auditory feedback is not ideal for applications involving navigation or general entertainment.

Given this context, tactile feedback may have many advantages. First, the sense of touch is very infrequently used for interactions. So, sending additional information via this channel will have little or no effect on the workload [LOT 13], and so will not affect performance. Second, unlike visual and auditory feedback, tactile feedback is personal, and is not perceived by others in the user's immediate environment. Motivated by this, various different types of tactile feedback were tested with BCIs. For example, the feedback can be sent in the form of vibration whose frequency changes as a function of the recognized MI task [CHA 07]. Using a single vibrotactile stimulator on the biceps, it is possible to feel whether the recognized task is an imagined movement of the left hand or the right hand. This study also shows that the performance for a given MI task (MI left hand or right hand) improves when the motor is placed on the same side as the task. This result throws back to the theory of *control-display mapping* [THU 12], which states that the effectiveness of tactile feedback depends on its coherency with the recognized MI task (e.g. stimulate the right hand when a MI task of the right hand is recognized). Other studies [KAU 06, CIN 07, CHA 12, LEE 13] suggest using multiple stimulators to provide feedback to the user. These studies focus in particular on applications with disabled users. The stimulators are placed on the neck or higher back (where sensitivity is preserved). Various different stimulation patterns based on the principles of *control-display mapping* are used as feedback, such as variations in the intensity or the spatial localization. In a recent study [JEU 15b], we tested a *continuous* tactile feedback system for the first time (continuous in the sense that it was updated four times per second), comparing it to an equivalent visual feedback system. We used gloves equipped with five vibrotactile stimulators each: a vibration in the left (or right) hand indicated recognition of the task "imagine moving the left (or right) hand". Furthermore, since we wanted to evaluate the relevance and effectiveness of these two types of feedback (visual and tactile) in interactive contexts, we added a *gamified* task of counting visual distraction elements in the learning environment. Our results show that the performance (combined scores for motor imagery tasks and counting) of users receiving tactile feedback was higher than that of users receiving visual feedback. This study suggests that tactile feedback might be able to increase BCI performance, especially for interactive tasks. However, despite the fact the performance obtained with tactile feedback is often equivalent to that obtained with visual feedback, and sometimes even better [JEU 15b], and also generally provides a better user experience (in that it is considered more natural), the sense of touch has only been infrequently used for BCIs.

Finally, two other very specific types of feedback have been explored. Two studies [KAC 11, WIL 12] examined the application of electrotactile tongue stimulation. The tongue possesses receptors that allow an excellent resolution in space, and sensitivity is preserved even with damage to the spinal column. Two other studies [GOM 11, RAM 12] examined proprioceptive feedback (i.e. feedback that provides information about the position of different body parts and the force required to perform a movement) while operating a neuroprosthesis. These studies produced very good results, showing that proprioceptive feedback coupled with visual feedback produces an improvement in performance compared to only visual feedback. However, these methods are very expensive and invasive, and so are not suitable for general purpose applications.

In addition to using different senses, changes in the content of the feedback have also been investigated. For instance, Hwang *et al.* [HWA 09] suggested training based on neurofeedback. The feedback was represented in the form of a schematic map showing the various activated zones of the cortex in real time, which allowed users to improve their performance. Another study [KAU 11] shows that increasing the level of required attention by using multiple senses does not decrease the performance compared to traditional feedback. Although these approaches are promising, they have not yet been thoroughly explored.

Finally, some studies used a procedure that introduced a bias into the feedback (i.e. by leading users to believe that their performance was better than it actually was). For example, [BAR 10] showed that expert users were hindered by biased feedback, but that this procedure could sometimes prove useful to new users. Another result showed that uniquely positive feedback reduced the performance when used for a large number of sessions [KÜB 01]. These results suggest that the experience level of the user needs to be taken into account when designing the optimal feedback system.

11.4.4. *Learning environment*

Most types of feedback used with BCIs (in particular visual feedback systems) often trigger a decrease in user motivation and are generally associated with an average user experience. *Gamified* learning protocols were developed with the objective of maintaining motivation levels and improving the user experience. For example, McCraedie *et al.* [MCC 14] suggested two simple games based on the *ball–basket paradigm* (i.e. maneuvering a ball to pass through a basketball hoop) and the concept of a spaceship that must avoid asteroids. Other studies, summarized in a review by Lécuyer *et al.* [LÉC 08], even suggested gamified BCI learning protocols that integrated elements of virtual reality. In one of the games, the "use the force" application inspired by Star Wars allows users to levitate a spaceship by imagining moving their feet. Indeed, studies by Ron-Angevin and Díaz-Estrella [RON 09] and Leeb *et al.* [LEE 06] show that using fun protocols, in particular protocols based on virtual reality, an increase in performance is observed for controlling BCIs compared to traditional learning protocols. Although these protocols are effective, they all use feedback that is visual in nature. However, as we have seen, the visual channel is often overloaded in interactive situations for which BCIs might be useful. It would therefore certainly be productive to combine these learning environments with tactile feedback systems, as in the study by Jeunet *et al.* [JEU 15b], and then compare the performance with learning situations in traditional environments.

11.4.5. *In summary: guidelines for designing more effective training protocols*

In this section, we will provide a summary of the guidelines that arise from the studies presented above, the objective of which is to act as a guide for whoever wishes to implement more effective training protocols:

– *Instructions*: it appears to be necessary to explicitly specify the learning objective to the user, in particular the fact that the user must learn to generate a stable, specific signal when performing the different MI tasks in order to be able to control the BCI in the long term. Furthermore, it is important to allow users to experiment independently rather than imposing any particular strategy for performing the tasks. On the other hand, for motor imagery, it appears that kinesthetic motor imagery is more effective than visual motor imagery;

– *Training tasks*: providing tasks that are designed to include a progression (increasing difficulty) and that are adaptive (specific to each user) appears to facilitate the acquisition of BCI-related skills. Including free-access and asynchronous sessions and preparatory practice tasks (e.g. meditation) also seems to help;

– *Feedback*: even though this has not been formally shown in a study, visual feedback with emotional connotations (e.g. smiley faces) seems to increase user motivation levels and consequently their performance. However, visual feedback is not ideal in interactive situations. The same is true for auditory feedback, which does not appear to be truly beneficial except for patients suffering from locked-in syndrome. Tactile feedback is promising, so long as the principles of *control-display mapping* are observed. Indeed, tactile feedback generally produces a level of performance equivalent to visual feedback, but relies on a channel that is much less saturated in interactive situations. Finally, increasing the quantity and the quality of the information provided (e.g. topography of cerebral activity) seems to be useful, as well as adapting the way that the feedback is presented to the experience level of the user;

– *Learning environment*: several studies have shown that gamified learning, especially including elements of virtual reality, increases the user motivation, and consequently performance.

11.5. Perspectives: toward user-*adapted* and user-*adaptable* learning protocols

The previous section presents the work performed until now toward the goal of improving BCI learning protocols. Indeed, we saw that certain studies attempted to build motivation in the learning environment (using games and virtual reality), but that improving the quality of the instructions and the relevancy of the practice tasks and the feedback provided to the user is also important. This review of the literature allows us to observe that the large majority of the work performed on improving BCI training protocols is situated within a larger trend that may be described as *technocentric*. In other words, the objective is to improve the BCI performance by modifying the learning protocols, but all efforts are concentrated on technological aspects (improving the interface and the content of the information provided to the user during learning). In this last section, we will present an emerging perspective in BCIs that may be described as *anthropocentric*, which breaks with but also complements the traditional approach presented above. It is based on a perspective of learning according to which individuals possess personal characteristics, inherited from past experiences, from their culture and thus encoded into their cognitive profile, their personality, etc. Improving

BCI learning therefore occurs by adapting the learning protocols to the personal characteristics of each learner.

This perspective came about from the observation that there is a large amount of variability in BCI studies in terms of performance. It is often the case that with the same learning protocol, certain subjects only achieve performances of approximately the random success rate, whereas others achieve almost 100% success. The question of *why do certain subjects succeed in learning when others fail?* is a legitimate question. This question recently inspired a study [JEU 15a] during which (1) participants were asked to learn to use a BCI based on performing 3 mental tasks (imagine moving the left hand, mental rotation and mental subtraction) [FRI 13] using a traditional protocol [PFU 01] over six sessions (performed over the course of 6 days) and (2) certain aspects of their cognitive profiles and personality were assessed using a selection of different questionnaires. This study produced two major results. The first result is that there is a strong correlation between a *mental rotation* trial score [VAN 78] and the average user performance at MI tasks. This initial observation suggests many interesting perspectives for BCI learning. Indeed, it is possible to imagine learning protocols during which the users' spatial skills are progressively improved, until they can successfully perform MI tasks. Second, the results of the study suggest that it may be possible to establish a model that allows the BCI performance to be predicted from three personality factors: (1) the stress or anxiety level of the user, (2) the user's abstraction/imagination skills and (3) the user's level of autonomy within a group. Similarly to the previous result, this approach provides a framework for a number of promising perspectives for implementing learning protocols adapted to the personal characteristics of the user.

This idea that learning protocols might be adapted to the characteristics of the learner leads us to consider solutions that were developed in the domain of *intelligent tutoring systems* (ITSs) [NKA 10]. ITSs are adaptive computer systems whose purpose is to support the learning of certain concepts by adapting the protocol to the user. An ITS provides sequences of exercises that allow the user to acquire a certain skill. The benefit of ITSs is that (1) this sequence of exercises is *adapted to the profile of the learner* and (2) this sequence is *adaptable during the learning process to match the state of the learner*. Before the learning sequence is started, the system first determines the learner's profile in order to provide a suitably adapted learning protocol (for example determining whether the learner is more *visual* or *verbal*, and providing exercises that are either picture-based or text-based accordingly). The learning process also adjusts to account for the progression of the learner's skill level (i.e. the learner's cognitive state). An increasing amount of effort has been invested in evaluating the progression of learners' emotional and mental states during the learning process. Certain emotions (such as enthusiasm or disappointment) are considered "academic emotions" [PEK 02], as they play a role in the success (or failure) of the learning process. Various indicators can be used to measure these states: behavioral (mouse movements, posture, facial expressions), physiological (variations in heart rate, electrodermal response or breathing) and neurophysiological (modifications of certain cerebral rhythms, measured by EEG). All of these states are objectified with the goal of adapting the exercise sequence in real time, and thus improving the efficiency of the learning process. For example, if an evaluation of the cognitive state finds that certain skills have not been acquired, the system will return to exercises that are designed to reinforce these skills. Similarly, if a decrease in the learner's motivation or enthusiasm is observed and the skills appear to have been acquired, the difficulty of the exercises can be increased more rapidly than was initially planned. ITS-supported learning (just like BCI learning) is a remote learning process: the learner is alone in front of a computer. But it is an established fact that *social presence* is an important factor in improving the efficiency of a learning process [GUN 95]. It has been shown that learning in the presence of a teacher, or other people, even passive observers, leads to an improvement in performance. This is the main reason why *virtual learning companions* (VLCs) were developed in connection with ITS. These VLC, often displayed on the ITS interface, can be configured to provide different kinds of feedback (social presence, emotional or cognitive support) depending on the profile and the state of the learner.

The idea of an ITS specifically designed for BCI learning seems promising, but raises a series of questions. For example, learning mathematics involves determining a sequence of competencies to be acquired, ranging from simple competencies (adding numbers) to more elaborate ones (performing operations on fractions). Defining a sequence of competencies to be acquired in BCI learning has proven to be much more complex. Another unique aspect of BCI learning is that learners are immobile during the learning process, which limits the number of behavioral indicators available for judging the learners' emotional and motivational states. Although these questions still require answers, we can easily imagine a simplified version of an ITS for BCI learning composed of several stages. First, the learner's profile must be characterized. To achieve this, the factors that we know *a priori* are involved in BCI performance (i.e. levels of stress, autonomy, imagination) must be evaluated. A VLC could then provide feedback that is adapted to this profile. For example, it could provide emotional support to anxious subjects, or a social presence to subjects with low "autonomy" scores. Next, using the participants' scores on the Mental Rotation trial, a sequence of exercises of progressively increasing difficulty could allow users to learn to improve their spatial skills, which, together with the support provided by the VLC, should have positive consequences on their BCI performance.

11.6. Conclusions

This chapter gives an idea of the current state of research of BCI learning protocols. The BCI community now recognizes that in order to achieve an improvement in performance, the user must be included in the loop, and so learning protocols must be improved accordingly. We have seen that a few promising avenues regarding the various constituent elements of these learning protocols (instructions, training tasks, feedback and learning environments) have been explored. Unfortunately, these types of study remain few and far between and, critically, their results are insufficiently utilized by the BCI community. We have also shown that by building on theories in disciplines such as the psychology of learning, it is possible to suggest new, promising approaches for improving user performance. One of the most important steps seems to be making the effort of understanding how each user works cognitively in order to offer learning protocols adapted to their individual profiles.

11.7. Bibliography

[ALL 10] ALLISON B., NEUPER C., *Could Anyone Use a BCI?*, Springer, London, 2010.

[BAR 10] BARBERO A., GROSSE-WENTRUP M., "Biased feedback in brain–computer interfaces", *Journal of Neuroengineering and Rehabilitation*, vol. 7, p. 34, 2010.

[BIR 99] BIRBAUMER N., GHANAYIM N., HINTERBERGER T. *et al.*, "A spelling device for the paralysed", *Nature*, vol. 398, pp. 297–298, 1999.

[CHA 07] CHATTERJEE A., AGGARWAL V., RAMOS A. *et al.*, "A brain–computer interface with vibrotactile biofeedback for haptic information,", *Journal of Neuroengineering and Rehabilitation*, vol. 4, p. 40, 2007.

[CHA 12] CHAVARRIAGA R., PERRIN X., SIEGWART R. *et al.*, "Anticipation- and error-related EEG signals during realistic human-machine interaction: a study on visual and tactile feedback", *IEEE Engeneering in Medicine and Biology Society*, pp. 6723–6726, 2012.

[CIN 07] CINCOTTI F., KAUHANEN L., ALOISE F. *et al.*, "Vibrotactile feedback for brain–computer interface operation", *Computational Intelligence and Neuroscience*, vol. 2007, p. 12, 2007.

[ESK 08] ESKANDARI P., ERFANIAN A., "Improving the performance of brain–computer interface through meditation practicing", *30th Annual International Conference of the IEEE Engineering in Medicine and Biology Society, (EMBS'08)*, IEEE, pp. 662–665, 2008.

[FRI 13] FRIEDRICH E.V., NEUPER C., SCHERER R., "Whatever works: a systematic user-centered training protocol to optimize brain–computer interfacing individually,", *PLoS One*, vol. 8, no. 9, p. e76214, 2013.

[GAR 12] GARGIULO G.D., MOHAMED A., MCEWAN A.L. *et al.*, "Investigating the role of combined acoustic-visual feedback in one-dimensional synchronous brain computer interfaces, a preliminary study", *Medical Devices: Evidence and Research*, vol. 5, pp. 81–88, 2012.

[GOM 11] GOMEZ-RODRIGUEZ M., PETERS J., HILL J. *et al.*, "Closing the sensorimotor loop: haptic feedback facilitates decoding of motor imagery", *Journal of Neural Engineering*, vol. 8, no. 3, pp. 36–45, 2011.

[GUN 95] GUNAWARDENA C.N., "Social presence theory and implications for interaction and collaborative learning in computer conferences", *International Journal of Educational Telecommunications*, vol. 1, pp. 147–166, 1995.

[HIN 04] HINTERBERGER T., NEUMANN N., PHAM M. *et al.*, "A multimodal brain-based feedback and communication system", *Experimental Brain Research*, vol. 154, pp. 521–526, 2004.

[HWA 09] HWANG H.-J., KWON K., IM C.-H., "Neurofeedback-based motor imagery training for brain–computer interface (BCI)", *Journal of Neuroscience Methods*, vol. 179, pp. 150–156, 2009.

[JEU 14] JEUNET C., CELLARD A., SUBRAMANIAN S. *et al.*, "How well can we learn with standard BCI training approaches? A pilot study", *6th International Brain–Computer Interface Conference*, 2014.

[JEU 15a] JEUNET C., N'KAOUA B., SUBRAMANIAN S. *et al.*, "Predicting mental imagery-based BCI performance from personality, cognitive profile and neurophysiological patterns", *PLoS One*, vol. 10, no. 10, p. e0143962, 2015.

[JEU 15b] JEUNET C., VI C., SPELMEZAN D. *et al.*, "Continuous tactile feedback for motor-imagery based brain–computer interaction in a multitasking context", *Human-Computer Interaction–INTERACT'15*, Springer, International Publishing, pp. 488–505, 2015.

[KAC 11] KACZMAREK K.A., "The tongue display unit (TDU) for electrotactile spatiotemportal parttern presentation", *Scientia Iranica*, vol. 18, no. 6, pp. 1476–1485, 2011.

[KAU 06] KAUHANEN L., PALOMÄKI T., JYLÄNKI P. *et al.*, "Haptic feedback compared with visual feedback for BCI", *Proceedings of the 3rd International Brain–Computer Interface Workshop & Training Course*, 2006.

[KAU 11] KAUFMANN T., WILLIAMSON J., HAMMER E. *et al.*, "Visually multimodal vs. classic unimodal feedback approach for SMR-BCIs: a comparison study", *International Journal of Bioelectromagnetism*, vol. 13, pp. 80–81, 2011.

[KOB 13] KOBER S.E., WITTE M., NINAUS M. *et al.*, "Learning to modulate one's own brain activity: the effect of spontaneous mental strategies", *Frontiers in Human Neuroscience*, vol. 7, 2013.

[KÜB 01] KÜBLER A., NEUMANN N., KAISER J. *et al.*, "Brain–computer communication: self-regulation of slow cortical potentials for verbal communication", *Archives of Physical Medicine and Rehabilitation*, vol. 82, pp. 1533–1539, 2001.

[LEE 06] LEEB R., KEINRATH C., FRIEDMAN D. *et al.*, "Walking by thinking: the brainwaves are crucial, not the muscles!", *Presence: Teleoperators and Virtual Environments*, vol. 15, pp. 500–514, 2006.

[LEE 07] LEEB R., LEE F., KEINRATH C. *et al.*, "Brain–computer communication: motivation, aim, and impact of exploring a virtual apartment", *IEEE Transactions on Neural Systems and Rehabilitation Engineering*, vol. 15, no. 4, pp. 473–482, 2007.

[LEE 13] LEEB R., GWAK K., KIM D.-S. *et al.*, "Freeing the visual channel by exploiting vibrotactile brain–computer interface feedback", *IEEE Engineering in Medicine and Biology Society*, vol. 2013, pp. 3093–3096, 2013.

[LOT 13] LOTTE F., LARRUE F., MÜHL C., "Flaws in current human training protocols for spontaneous BCI: lessons learned from instructional design", *Frontiers in Human Neurosciences*, vol. 7, p. 568, 2013.

[LÉC 08] LÉCUYER A., LOTTE F., REILLY R.B. *et al.*, "Brain–computer interfaces, virtual reality, and videogames", *IEEE Computer*, vol. 41, no. 10, pp. 66–72, 2008.

[MCC 14] MCCRAEDIE K.A., COYLE D.H., PRASAD G., "Is sensorimotor BCI performance influenced differently by mono, stereo, or 3-D auditory feedback?", *IEEE Transactions on Neural System and Rehabilation*, vol. 22, no. 3, pp. 431–440, 2014.

[MCF 10] MCFARLAND D., SARNACKI W., WOLPAW J., "Electroencephalographic control of three-dimensional movement", *Journal of Neural Engineering*, vol. 7, no. 3, 2010.

[NEU 03] NEUPER C., MÜLLER G., KÜBLER A., BIRBAUMER N. *et al.*, "Clinical application of an EEG-based BCI: a case study in a patient with severe motor impairment", *Clinical Neurophysiology*, vol. 114, no. 3, pp. 399–409, 2003.

[NEU 05] NEUPER C., SCHERER R., REINER M. *et al.*, "Imagery of motor actions: differential effects of kinesthetic and visual–motor mode of imagery in single-trial EEG", *Cognitive Brain Research*, vol. 25, pp. 668–677, 2005.

[NEU 10] NEUPER C., PFURTSCHELLER G., "Neurofeedback training for BCI control", *Brain-Computer Interfaces*, Springer, Berlin Heidelberg, 2010.

[NIJ 08] NIJBOER F., FURDEA A., GUNST I. *et al.*, "An auditory brain–computer interface (BCI)", *Journal of Neuroscience Methods*, vol. 167, pp. 43–50, 2008.

[NKA 10] NKAMBOU R., BOURDEAU J., MIZOGUCHI R., *Advances in Intelligent Tutoring Systems*, Springer, 2010.

[PEK 02] PEKRUN R., GOETZ T., TITZ W. *et al.*, "Academic emotions in students' self-regulated learning and achievement: a program of qualitative and quantitative research", *Educational Psychologist*, vol. 37, pp. 91–105, 2002.

[PFU 01] PFURTSCHELLER G., NEUPER C., "Motor imagery and direct brain–computer communication", *Proceedings of the IEEE*, vol. 89, no. 7, pp. 1123–1134, 2001.

[RAM 12] RAMOS-MURGUIALDAY A., SCHÜRHOLZ M., CAGGIANO V. *et al.*, "Proprioceptive feedback and brain computer interface based neuroprostheses", *PLoS One*, vol. 7, no. 10, 2012.

[RON 09] RON-ANGEVIN R., DÍAZ-ESTRELLA A., "BCI: changes in performance using virtual reality techniques", *Neuroscience Letters*, vol. 449, pp. 123–127, 2009.

[SMI 05] SMITH E., DELARGY M., "Locked-in syndrome", *BMJ*, vol. 330, pp. 406–409, 2005.

[THU 12] THURLINGS M.-E., VAN ERP J.B., BROUWER A.-M. *et al.*, "Control-display mapping in brain–computer interfaces", *Ergonomics*, vol. 55, no. 5, pp. 564–580, 2012.

[VAN 78] VANDENBERG S.G., KUSE A.R., "Mental rotations, a group test of three-dimensional spatial visualization", *Perceptual and Motor Skills*, vol. 47, pp. 599–604, 1978.

[VID 10] VIDAURRE C., BLANKERTZ B., "Towards a cure for BCI illiteracy", *Brain Topography*, vol. 23, pp. 194–198, 2010.

[WIL 12] WILSON J.A., WALTON L.M., TYLER M. *et al.*, "Lingual electrotactile stimulation as an alternative sensory feedback pathway for brain–computer interface applications.", *Journal of Neural Engineering*, vol. 9, no.4, p. 045007, 2012.

[WOL 00] WOLPAW J., MCFARLAND D., VAUGHAN T., "Brain–computer interface research at the Wadsworth Center", *IEEE Transactions on Rehabilitation Engineering*, vol. 8, no. 2, pp. 222–226, 2000.

[WOL 02] WOLPAW J.R., BIRBAUMER N., MCFARLAND D.J. *et al.*, "Brain–computer interfaces for communication and control", *Clinical Neurophysiology*, vol. 113, pp. 767–779, 2002.

12

Brain–Computer Interfaces for Human–Computer Interaction

Research on BCIs has so far usually focused on processing and classifying brain signals with the objective of improving the speed and precision of the interface. Progress in these areas has allowed us to diversify the applications of BCIs. It is now time to work on improving the interactions that occur through these interfaces.

We will study in this chapter the relationship between BCIs and Human–Computer Interaction (HCI), and we will see how our knowledge in HCI can be applied to BCIs. We will first begin with a general overview of the principal concepts of HCI. We will then study the most important properties of BCIs in terms of these concepts. This chapter will also discuss the problem of choosing the right cerebral pattern for a given interaction and usage context. Finally, we will present the most promising recent paradigms of BCI interaction.

12.1. A brief introduction to human–computer interaction

Human-computer interaction (HCI) is a discipline concerned with the design, evaluation and implementation of interactive computing systems for human use [HEW 92]. In this section, we will simply define a few important

Chapter written by Andéol EVAIN, Nicolas ROUSSEL, Gry CASIEZ, Fernando ARGELAGUET-SANZ and Anatole LÉCUYER.

concepts from this vast field of research. Interested readers can refer to [JAC 12] for a much more complete introduction to the subject.

12.1.1. *Interactive systems, interface and interaction*

An *interactive system* is a system whose operations depend on an unpredictable input from an external environment that it does not control [GOL 06]. The *interface* is the set of hardware and software mechanisms that allow a person to operate, control and supervise an interactive computer system. *Interaction* occurs between the user and the system. This is the object of study in HCI, which aims to understand it (i.e. observe it, describe it, explain it) and improve it.

12.1.2. *Elementary tasks and interaction techniques*

The operation and control of an interactive computer system are founded on a set of *elementary tasks* that the user can achieve. Each task can be performed by means of a set of various *interaction techniques*. The elementary tasks are the smallest units of operation possible in a given context. An interaction technique is a certain combination of hardware and software mechanisms that accomplishes a given task. The task represents part of the objective, whereas the technique represents part of the means by which that objective is achieved.

Elementary interaction tasks vary in nature according to the domain of application. For example, Foley *et al.* list six elementary tasks for graphical interactions: selecting, positioning, orienting, tracing, quantifying, and text input [FOL 84]. Touching an object on a touch screen or indirectly specifying it by clicking on it with a mouse are two examples of interaction techniques for selecting that object. Operating a physical potentiometer, a virtual potentiometer with the mouse or entering text are three possible techniques for specifying a numerical value. Voice recognition or keyboard input are two possible techniques for entering text.

12.1.3. *Theory of action feedback*

The *theory of action* outlined by Norman deconstructs the act of performing a task into seven stages: establishing the goal, forming the intention, specifying the action sequence, executing the action, perceiving the system state, interpreting the state, and evaluating the system state with respect to the goals and intentions [NOR 86] (see Figure 12.1). These seven stages are not all necessarily present, and can occur in a different order, but this decomposition is nonetheless useful for analyzing and designing interactive systems.

The user's mental picture of certain concepts might be very different from the way that these concepts are implemented by the system. For Norman, there are two gulfs separating the user's conceptions from the system's conceptions: the gulf of execution and the gulf of evaluation. The terms *distance of execution* and *distance of evaluation* describe the effort that the user or the system designer must invest in order to cross these gulfs. The adjectives *semantic* and *articulatory* are used to distinguish efforts related to the meaning of user-system exchanges from efforts related to the form that these exchanges take.

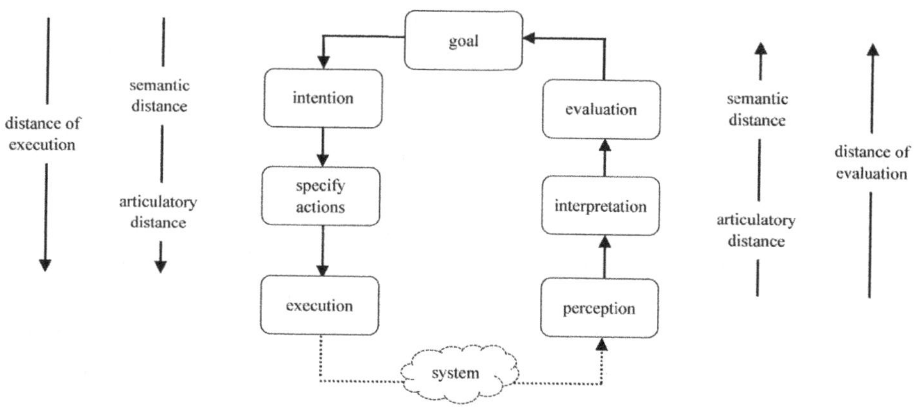

Figure 12.1. *The seven stages of user activity when performing a task and the corresponding distances [NOR 86]*

The speed and the form of the *feedback* produced by the system largely determine the capacity of users to perceive, interpret and evaluate their state changes, and thus affect the evaluation distance. For instance, prompt feedback builds a continuous representation of the state of the system, and the effect of actions as they are performed. Prompt feedback also contributes to the sensation that the user is acting directly upon the objects of interest, allowing the user to feel engaged in the task.

Interaction techniques are usually designed around the requirements of the tasks, leveraging as fully as possible the users' cognitive, motor and perceptual skills to reduce the execution and evaluation distances. But no matter how carefully the technique is designed, an interaction technique that is perfectly adapted for one task in a certain context may prove unsuitable in others. Text input by voice recognition is undoubtedly preferable to using a keyboard while driving, for example. The question of whether a certain interaction technique is suitable for a certain task in a certain context is the question of *usability*.

12.1.4. *Usability*

Usability is defined by the ISO 9241 standard as *the extent to which a product can be used by specified users to achieve specified goals with effectiveness, efficiency, and satisfaction in a specified context of use* [ISO 98]. Effectiveness refers to the capacity of attaining specified goals, efficiency describes the resources spent in order to achieve these goals and satisfaction describes how the user perceives the process.

Usability may be evaluated using a variety of different criteria. One might wish that inexperienced users find a system easy to use. Relevant indicators in this case might, for example, include the percentage of tasks successfully performed on the first attempt, the amount of time required to do so and the proportion of deliberately performed actions. Alternatively, one might wish a system that is robust to user mistakes. In this case, the percentage of errors corrected by the system, the time taken to do so and the user's appraisal of the corrections made are of interest. Another possible choice is a system that is easy to learn. We would then look at the number of functions acquired, the time taken to do so and the user's appraisal of the training process. The choice of usability criteria naturally depends on the domain of application.

12.2. Properties of BCIs from the perspective of HCI

BCIs measure cerebral activity signals, filter these signals, extract features from them and then classify the vector of features thus obtained. The class that is determined at the end of this chain of processes is then used to activate or configure an interaction technique. The characteristics of the interaction controlled by BCI vary greatly depending on the chosen cerebral pattern. Certain interfaces use potentials triggered by external stimuli, such as P300 or SSVEP. These interfaces often achieve relatively high information transfer rates, but a long exposure to the flickering can be tiring. Conversely, cerebral patterns such as SCP or sensorimotor rhythms can be controlled by experienced users without external stimulation. For a more complete review of the most commonly used cerebral patterns with BCIs, interested readers can refer to Chapter 4.

Despite differences in their individual characteristics, there is a set of common properties shared by all BCIs in general, which may be compared with those of more standard interfaces. For example, the information transfer rate of a BCI is always significantly lower than the expected transfer rate achievable with a keyboard or a mouse. We will now present the most important of these aspects, common to all BCIs.

Latency is one of the more obvious properties of BCIs. The latency of most current interfaces is of the order of a few seconds. Indeed, the features of the signals must be measured over a minimum period that is sufficiently large that the cerebral patterns can be classified with reasonable levels of accuracy. For example, at a measured neuronal activation frequency of 5 Hz (i.e. the period of the signal is 200 ms), we cannot reasonably expect to obtain acceptable levels of precision in less than a second. The acceptability of latency in an interactive system depends on the sensory perception threshold of the user. A latency period of one second very distinctly exceeds the thresholds of auditory and visual perception. Fortunately, in the case of BCIs, the effective threshold is somewhat higher, as our perception of the point in time at which a specific mental state was produced is less precise than our perception of the point in time at which a button was pressed, for example. For comparison, interactive graphical systems controlled via mouse and keyboard typically have latencies of less than 100 ms between the physical action performed by the user and the display of updated information on the screen. The *precision* of a BCI when identifying a command is equally relatively low. Errors can arise from the user, who may not be able to adequately produce the required mental state, or, as is more commonly assumed, from the signal processing and classification steps. Most BCI research until now has focused on improving this precision, and significant progress has been made. Despite this progress, the precision still remains strongly user dependent, and for certain kinds of cerebral pattern the classification rate is still fairly low. For a BCI, a classification rate of 90% is considered to be good, whereas in normal conditions virtually every action performed with a mouse or keyboard is accurately registered by the peripheral.

The *number of commands* that may be accessed via a BCI is limited. Even for experienced users, the current levels of classifier precision do not generally allow for large numbers of classes without a drop in the detection rate. However, for some applications, having only three or four available commands does not necessarily represent a limitation. Note also that the P300 method makes it possible to choose from a large number of commands by successively selecting subsets without requiring the user to explicitly alter their mental state for each command.

The *information transfer rate* between the user and the computer through the BCI, largely limited by the properties of BCIs listed above, only exceeds 100 bpm in very rare cases [DON 00, WAN 08]. For comparison, the information transfer rate that an experienced user can achieve with a keyboard is of the order of 900 bpm (assuming a typing speed of 300 characters per minute and a Shannon entropy of the order of 3 bits/character). With a mouse, the information transfer rate is of the order of a few hundred bits per minute [MAC 92].

There is significant *variability in the performance* between different BCI users. Some users manage to operate BCIs much more effectively than the average, whereas others, sometimes called "BCI-illiterate", are completely incapable of using them. Even between users that can use BCIs effectively, large degrees of variability are observed from session to session.

BCIs do not require any *motor activity* from the user, as they take input directly from brain activity. They may therefore generally be used by individuals with motor handicaps. Furthermore, other channels of interaction, such as the hands, remain available for controlling other devices (hybrid approach). It can, however, be difficult to divide attention between multiple different tasks, even if adapted methods of feedback (e.g. tactile) can help [LEE 13a]. Muscle activity also represents a source of noise in the signal with potentially very high amplitude, which can make it more difficult to interpret. Blinking and jaw contractions are particularly problematic. These artifacts can be corrected using an EOG based method for "cleaning" the signal [SCH 07b]. The compatibility of BCIs with other interfaces such as a keyboard or mouse is currently a subject of research [MER 13, LEE 13b]. The *distance of execution* (see section 12.1.3) of BCIs is high, as the mental state associated with the desired task must be produced by the user. With practice, the association between accomplishing a goal and producing an intermediate mental task becomes more intuitive. The distance of execution may be reduced by choosing mental tasks that are similar to the physical tasks, e.g. imagining a movement of the left hand to move a cursor to the left.

The *hardware and software* used by BCIs is relatively complex to install and operate. This complexity may slow the propagation of BCIs in domestic environments (e.g. for video games), and is also the subject of research [DUV 12, FRE 14].

12.3. Which pattern for which task?

We saw in section 12.1.2 that any interactive task may be decomposed into elementary tasks. Depending on the nature of the elementary task, certain cerebral patterns are more suitable than others; choosing appropriately allows the distance of execution to be further reduced.

Text entry involves entering sequences of characters into the system. It is typically achieved with a keyboard. This task was one of the first tasks to be realized using BCIs, usually with the P300 [LOT 08], motor imagery [BLA 06] or SSVEP [WAN 08]. The performance achieved is in the region of seven characters per minute [DON 00], compared to 20–40 using a keyboard for a non-expert user.

Quantification involves specifying a numerical value between some maximal and minimal thresholds. BCIs have not often been used for this task, but producing motor imagery patterns with specific or high amplitude levels has been used as a challenge in video games [LÉC 13, HJE 03].

Selection involves choosing one or multiple elements from a set of fixed size (e.g. menus, radio buttons, checkboxes) or of variable size (e.g. 2D or 3D targeting, dropdown boxes, selectboxes). P300 allows users to select one element from a few dozen, striking a good compromise between speed and precision, e.g. one element from 36 in 7.7 s, with an 80% success rate [DON 00]. SSVEP also allows users to perform selection tasks. For example, one target can be selected out of six in 1 second, with a precision of 86.7% [WAN 08]. For selection by targeting, it is possible to control two combinations of brain rhythms in the sensorimotor regions to control a cursor and select or reject a target highlighted using a third combination [VAU 06]. This technique however requires a large degree of concentration from the user, and a significant amount of practice, e.g. 5–15 h divided into sessions of 24 min over several weeks [MCF 08].

Manipulation and transformation involve modifying the position, orientation, size or shape of an object. In the case where the set of possible manipulations and transformations is small and discrete, these tasks may be performed with a selection task that chooses the desired modification, and so BCIs can be used (see previous section) [LEG 13].

Navigation involves changing the viewpoint in a virtual setting, for example by repositioning the camera in a 3D environment, or scrolling through the content of a document. An indirect form of navigation with a BCI can be achieved by specifying high-level commands. For example, the destination can be selected using a P300, then transmitted to an automatic navigation system [REB 07]. Motor imagery can provide more direct and refined navigation with low distances of execution. For example, the user can imagine moving the left or right hand to turn in that direction [LEE 07].

In addition to the classical tasks listed above, BCIs can also be used to evaluate the user's mental workload and detect certain emotional states such as wakefulness, pleasure, drowsiness, prevalence or frustration [HER 07, GEO 12]. Few other systems offer this range of possibilities. Motion tracking and physiological sensors that, for example, measure the galvanic skin response or cardiac rhythm are the main alternatives.

12.4. Paradigms of interaction for BCIs

12.4.1. *BCI interaction loop*

To integrate BCIs into applications, the exchange of information between each component of the interaction must be clearly defined. The simplest way to consider the role of the BCI within the structure of a wider interaction is to see it as an external component, as presented in [HIN 13]. The signal processing chain is the centerpiece of the BCI. The acquisition system registers a signal derived from brain activity. This signal is then filtered and transmitted to the feature extraction block. The extraction block calculates the values of the desired features. The resulting features vector must then be classified. This function is performed by the classification block, which then sends the results to the application. The application is responsible for associating the detected cerebral patterns with the command to be performed. We will now extend this conceptual model of a BCI to account for feedback, and to make a distinction between the measured brain activity and conscious thought (see Figure 12.2).

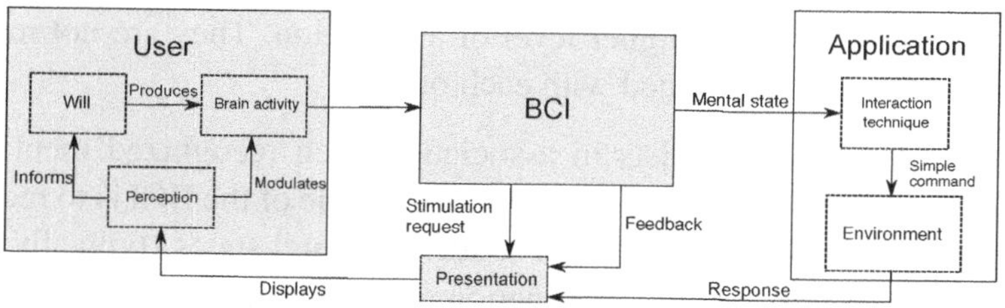

Figure 12.2. *Classical BCI interaction loop: Components of the BCI interaction loop: The user attempts to produce an intermediate mental state to interact with the machine through the BCI. Given the right stimulation, this mental state will produce a recognizable cerebral pattern for the BCI*

External *stimulation* (generally visual) is necessary for some cerebral patterns. The features that must be extracted to detect the cerebral pattern depend on the timing of this stimulation. Thus, the stimulation block must provide synchronization information to the feature extraction block. This block must also control the visual (or auditory, or tactile) display to the user. Finally, depending on the technique of interaction, it can be useful to adapt the stimulation as a function of the most recently detected cerebral patterns. Hence, the classification block also sends occasional output to the stimulation block.

The *presentaion* block is charged with gathering data and sending it to the user. This information includes neuronal stimulation, feedback and application-dependent data.

The user's *brain activity* is distinct from his or her will. Using a BCI requires the user to produce an intermediate mental state that must be controlled, much like how classical interfaces require an intermediate motor task. Ideally, any stimulation produced by the BCI should only influence the brain signal (measured by EEG), without affecting any conscious state, so that the stimulus is not irritating. In practice, if the amplitude of a certain instance of the brain activity is large enough to be picked up by EEG, it also draws the user's attention. Conversely, the application's response must access the user's conscious state, without overly interfering with the EEG signal. To be effective with BCIs, cerebral pattern must be insensitive to this type of noise.

12.4.2. *Main paradigms of interaction for BCIs*

For purposes of readability, the interaction paradigms outlined in this section are considered at a higher level of abstraction. They are not mutually exclusive and may be combined with each other:

– *Direct commands* consists in associating each recognized mental state with an explicit command (see Figure 12.2). The role of the BCI is to recognize the mental state from a finite set of classes of mental states (typically two or three classes) and to relay this information to the application. The application systematically relates each possible class to a command that will be executed when the mental state is detected.

Direct commands are probably the most widely used interaction paradigm for BCIs. Applications allow handicapped persons to operate computers or wheelchairs with direct commands [REB 07, VAU 06]. More recently, certain video games used the recognized mental state as the main input [LAL 05, NIJ 09]. Direct command BCIs are useful in situations where standard interactive devices are ineffective, e.g. because users require the use of their hands for some other task.

– The *hybrid approach* involves using the BCI as a complementary input device in combination with other devices (see Figure 12.3). Two or more interactive devices may be used simultaneously, each sending the information gathered to the application. The BCI is introduced at the same level as the other devices, and provides an additional information channel between the user and the machine. The other devices might, for example, be a keyboard, a mouse, a joystick [PFU 10] or another BCI [VAU 06]. Hybrid BCIs must be particularly insensitive to noise, because using other devices can generate additional artifacts due to eye movements and muscle contractions. The user must produce a cerebral pattern to control the BCI while simultaneously controlling the other devices as usual.

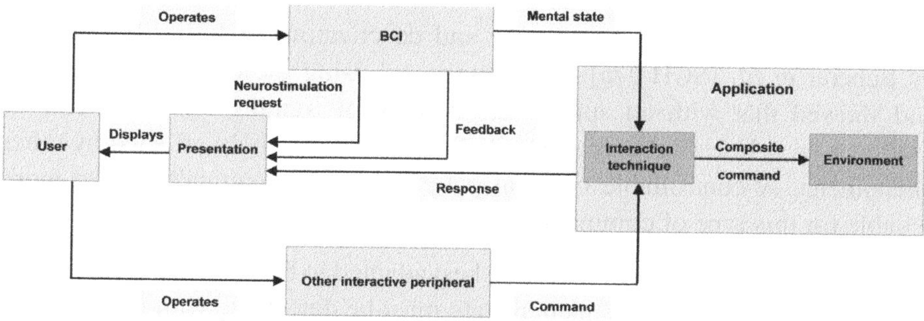

Figure 12.3. *Hybrid interaction with a BCI: one or more other complementary interaction devices are used simultaneously to accomplish a more complex interactive task*

Applications with hybrid BCIs use multiple inputs to improve the precision of the interface as a whole, or to specify different parameters of the same command [FRU 11, LI 10]. More recently, Zander *et al.* suggest using a gaze-tracking device in combination with a BCI to create a system capable of hands-free targeting and selecting that produces less false positives than traditional gaze-tracking systems, which are based on the eyes' fixation duration [ZAN 10]. As BCIs are used for increasingly complex tasks, the ability to use multiple inputs more independently could potentially become crucial. For example, in [LEE 13b], players can make their character run with a joystick and jump with a BCI.

– The *brain switch* involves using a BCI to activate or deactivate another interactive device. BCIs with a single command are sometimes also referred to as *brain switch* BCIs. From the perspective of the interaction, this is a special case of a direct command.

Two mental states are recognized by the brain switch: a "resting" state and an "action" state. Each time that the "action" state is detected, the other device is toggled. Thus, the brain switch may also be viewed as a special case of hybrid interaction. Indeed, the command to activate or deactivate the device can be transmitted to the application, which controls the device.

From the user's perspective, it is not necessary to concentrate on the BCI, except when using the specific activation command. When the user does this, he or she performs an intermediate mental task to produce the brain activity associated with the "action" command.

The need for a way of activating and deactivating BCIs has been raised by Scherer *et al.* [SCH 07a]. They suggested using the heartbeat to do this and showed that with an appropriate amount of training this input can be used as a switch. However, cardiac rhythms can be strongly affected by other phenomena. A controllable cerebral pattern might theoretically prove more reliable for this type of command [GEO 12].

The question of which patterns are best adapted to brain switch applications remains open. The "action" mental state must be detectable with a good level of precision and a low false positive rate (false positives lead to unwanted activations). On the other hand, since this command is only rarely used (typically once at the beginning of a session to activate the interface, and once at the end to deactivate it), it is acceptable for the activation period to be relatively high, with a large delay (approximately 30 s). Conservative approaches with classical cerebral patterns (SSVEP, P300) are good candidates for this interaction paradigm.

– The *passive BCI* approach involves detecting a mental state for purposes other than direct control (see Figure 12.4). The recognized cerebral patterns are sent to the application, which can use the user's mental state as an information parameter for adapting the main interaction technique [GEO 12]. From the user's perspective, there is no need to consciously control the cerebral patterns produced. Users can concentrate on their primary task rather than focusing attention on the BCI.

The "passive" approach has been used with other physiological markers such as the galvanic skin response [ALL 04] or gaze-tracking [HYR 06]. Cutrell and Tan suggested using BCIs for implicit interactions [CUT 08]. Since then, BCIs have been used to detect the level of engagement in a task, the user's mood, certain emotions, error recognition, and relaxation and mental workload [ZAN 11, GEO 12]. Detecting these kinds of mental state can be useful for adaptive automated processes (for example dynamic allocation of a task between the user and the machine), implicit markup of multimedia content, video games and error correction (see Chapter 5).

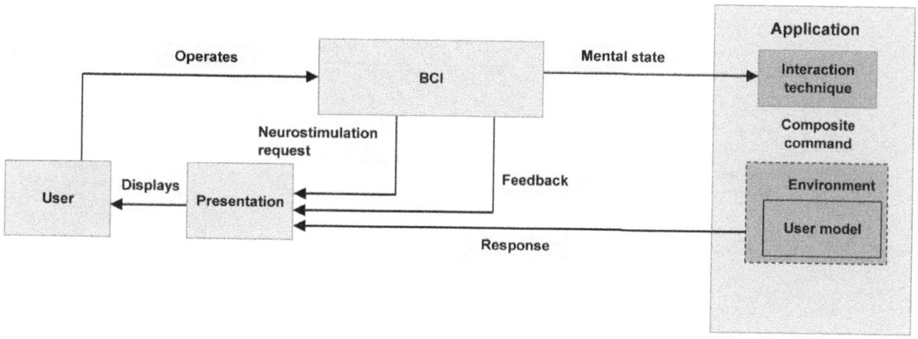

Figure 12.4. *Passive BCI: the user does not send an explicit command, but the application can construct a model of the user based on his or her mental state*

Passive BCIs can be useful in any kind of application in which the mental and emotional state of the user is relevant. From the machine's perspective, passive BCIs make it possible to dynamically adapt models of the user state.

– *Shared control* involves transforming the classes recognized by the BCI into high-level commands that are sent to the application (see Figure 12.5). Shared control is a paradigm of delegation. The machine is responsible for a certain fraction of the system's intelligence, in which high-level concepts and complementary information can be used to determine how a single brain command must be transformed into a more complex, high-level command or equivalently into several low-level commands.

From the user's perspective, the number of commands to be sent is low, even for accomplishing complex tasks. The quantity of information transmitted though the BCI is significantly reduced compared to direct command approaches. The user is therefore able to rest, while the high-level command is being executed [LOT 10].

It has been shown that shared control can be useful for operating wheelchairs [PHI 07]. For example, users can select a destination with a BCI based on the P300 paradigm, and a decision-making program equipped with a localization algorithm and object avoidance sensors decides the elementary actions that the wheelchair must perform as a result [REB 07].

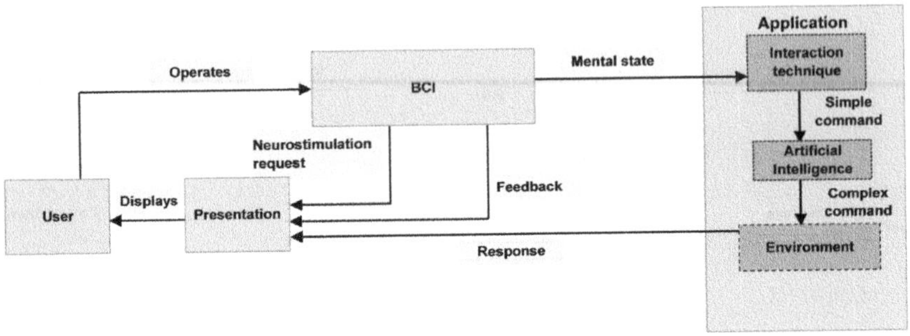

Figure 12.5. *Shared control: a unique command is analyzed and transformed into a high-level command that may be automatically deconstructed into a series of low-level commands*

Shared control can significantly speed up the interaction when the machine is capable of anticipating the user's decisions. As artificial intelligence and decision-making algorithms improve, shared control might find new fields of application.

– The *multi-user* approach involves using multiple BCI inputs to control a single application. Several mental activities may be combined at different stages of the interaction. The mental state of each user is recognized independently and sent to the application, which combines the recognized mental states in order to produce a command. The purpose of this combination might be to improve the global performance, or to increase the number of available commands. Mental activities can also be combined at the signal processing level, allowing the classifier itself to determine the multiuser command [BON 13]. Finally, multiuser approaches can be used to search for brain markers (*hyperscanning*) by observing the similarities between the brain activities of two users in similar situations [BAB 14].

An intermediate mental task must be performed by each user to create the correct mental state, and thus transmit the desired command, either in collaboration or in competition with the other user.

Recently, it was suggested that multiuser BCIs could be developed for applications in video games [NIJ 13]. Two players could attempt to synchronize their cerebral signals directly, or as a means of achieving higher level objectives. Competitive *gameplay* could also be introduced [BON 13].

Using the cerebral activity of multiple users could potentially improve the precision of BCIs, as the noise in each individual signal becomes less significant given the information from the other users. Concretely, the classifier could use features extracted from all users in a single classification step. The social presence of another user also appears to stimulate the learning process [RIC 03].

Future BCIs might belong to one of three possible usage categories. First, BCIs might be used as an alternative to the keyboard or the mouse. It is however not entirely clear that BCIs will be able to match the performance of these more classical tools. The nervous signal that directs muscular activity comes directly from the brain, whereas non-invasive technologies only have access to noisy signals. BCIs are disadvantaged at the outset by the clarity of the signal compared to traditional devices. However, direct and hybrid approaches might prove useful in assisting the interactions of handicapped individuals. For applications intended for wider audiences, BCIs might be used as complementary input devices that allow the usual channels of interaction to be kept free for other interfaces. Finally, even if other tools of interaction are more effective at accomplishing a given task, performing the task with a BCI might provide other advantages.

12.5. Conclusions

In this chapter, we saw that the cerebral patterns recognized by various different BCIs strongly affect the properties of the BCIs in terms of HCI. Nonetheless, there are many similarities (latency, precision) between BCIs and a wider set of interfaces. These similarities can limit the usefulness of BCIs for certain tasks, but are not necessarily relevant for others.

The right choice of usability criterion can vary wildly depending on the application, and the choice of a suitable cerebral pattern for controlling the BCI must be adapted to the final usage context. The classical BCI context, consisting of laboratory conditions with a single subject without distractions or any other tools, although not entirely obsolete, is too restrictive to properly describe future applications.

Until now, most research has focused on improving the precision and speed of BCIs, but in order to extend their range of applications, it is crucial that we also consider other usability criteria.

The unique qualities of BCIs make them useful for assisting handicapped individuals in situations where other interactive devices are ineffective. More recently, other fields of application have begun to emerge. In particular, BCIs might play an important role in future video games.

In order to meet the requirements of these new domains, the subject's ability to produce suitable cerebral patterns must be explored, as well as learning techniques for acquiring these kinds of skills. In parallel, the need for new cerebral patterns that can be exploited by BCIs will likely intensify. In all likelihood, we have only discovered a fraction of the possibilities. Advancements in neuroscience research might reveal new prospects.

12.6. Bibliography

[ALL 04] ALLANSON J., FAIRCLOUGH S.H., "A research agenda for physiological computing", *Interacting with Computers*, vol. 16, no. 5, pp. 857–878, 2004.

[BAB 14] BABILONI F., ASTOLFI L., "Social neuroscience and hyperscanning techniques: past, present and future", *Neuroscience & Biobehavioral Reviews*, vol. 44, pp. 76–93, 2014.

[BLA 06] BLANKERTZ B., DORNHEGE G., KRAULEDAT M. *et al.*, "The Berlin brain–computer interface: EEG-based communication without subject training", *IEEE Transactions on Neural Systems and Rehabilitation Engineering*, vol. 14, no. 2, pp. 147–152, 2006.

[BON 13] BONNET L., LOTTE F., LÉCUYER A., "Two brains, one game: design and evaluation of a multi-user BCI video game based on motor imagery", *IEEE Transactions on Computational Intelligence and AI in Games*, vol. 5, no. 2, pp. 185–198, 2013.

[CUT 08] CUTRELL E., TAN D., "BCI for passive input in HCI", *Proceedings of CHI*, vol. 8, pp. 1–3, 2008.

[DON 00] DONCHIN E., SPENCER K., WIJESINGHE R., "The mental prosthesis: assessing the speed of a P300-based brain–computer interface", *IEEE Transactions on Rehabilitation Engineering*, vol. 8, no. 2, pp. 174–179, 2000.

[DUV 12] DUVINAGE M., CASTERMANS T., DUTOIT T. *et al.*, "A P300-based quantitative comparison between the Emotiv Epoc headset and a medical EEG device", *Biomedical Engineering*, vol. 765, pp. 2012–764, 2012.

[FOL 84] FOLEY J., WALLACE V., CHAN P., "The human factors of computer graphics interaction techniques", *IEEE Computer Graphics and Applications*, vol. 4, no. 11, pp. 13–48, 1984.

[FRE 14] FREY J., GERVAIS R., FLECK S. *et al.*, "Teegi: Tangible EEG Interface", *UIST-ACM User Interface Software and Technology Symposium*, pp. 301–308, 2014.

[FRU 11] FRUITET J., CLERC M., PAPADOPOULO T., "Preliminary study for an offline hybrid BCI using sensorimotor rhythms and beta rebound", *International Journal of Bioelectromagnetism*, vol. 13, no. 2, pp. 70–71, 2011.

[GEO 12] GEORGE L., Contribution to the study of active and passive Brain–Computer interfaces for interacting with virtual environments based on EEG and mental workload, PhD Thesis, INSA Rennes, 2012.

[GOL 06] GOLDIN D., SMOLKA S.A., WEGNER P., *Interactive Computation: The New Paradigm*, Springer-Verlag, New York, 2006.

[HER 07] HERAZ A., FRASSON C., "Predicting the three major dimensions of the learner's emotions from brainwaves", *International Journal of Computer Science*, vol. 25, pp. 323–329, 2007.

[HEW 92] HEWETT T.T., BAECKER R., CARD S. *et al.*, *ACM SIGCHI Curricula for Human-Computer Interaction*, ACM, 1992.

[HIN 13] HINTERMÜLLER C., KAPELLER C., EDLINGER G. *et al.*, "Brain–Computer Interfaces systems: recent progress and future prospects", in *BCI Integration: Application Interfaces*, inTech, 2013.

[HJE 03] HJELM S.I., "The making of BrainBall", *Interactions*, vol. 10, pp. 26–34, 2003.

[HYR 06] HYRSKYKARI A., Eyes in attentive interfaces: experiences from creating iDict, a gaze-aware reading aid, PhD Thesis, University of Tampere, 2006.

[ISO 98] ISO, Ergonomic requirements for office work with visual display terminals – Part 11: guidance on usability, International standard no. 9241–11, ISO/TC 159, March 1998.

[JAC 12] JACKO J.A., (ed.), *The Human-Computer Interaction Handbook*, CRC Press, 2012.

[LAL 05] LALOR E.C., KELLY S.P., FINUCANE C. *et al.*, "Steady-state VEP-based Brain–Computer Interface control in an immersive 3D gaming environment", *EURASIP J. Appl. Signal Process.*, pp. 3156–3164, 2005.

[LÉC 13] LÉCUYER A., GEORGE L., MARCHAL M., "Toward adaptive VR simulators combining visual, haptic, and Brain–Computer Interfaces", *IEEE Computer Graphics and Applications*, vol. 33, no. 5, pp. 18–23, 2013.

[LEE 07] LEEB R., LEE F., KEINRATH C. *et al.*, "Brain–computer communication: motivation, aim, and impact of exploring a virtual apartment", *IEEE Transactions on Neural Systems and Rehabilitation Engineering*, vol. 15, pp. 473–482, 2007.

[LEE 13a] LEEB R., GWAK K., KIM D.-S. *et al.*, "Freeing the visual channel by exploiting vibrotactile BCI feedback", *Engineering in Medicine and Biology Society*, pp. 3093–3096, 2013.

[LEE 13b] LEEB R., LANCELLE M., KAISER V. *et al.*, "Thinking penguin: multimodal Brain–Computer Interface control of a VR game", *IEEE Transactions on Computational Intelligence and AI in Games*, vol. 5, no. 2, pp. 117–128, 2013.

[LEG 13] LEGENY J., VICIANA-ABAD R., LECUYER A., "Toward contextual SSVEP-based BCI controller: smart activation of stimuli and control weighting", *IEEE Transactions on Computational Intelligence and AI in Games*, vol. 5, pp. 111–116, 2013.

[LI 10] LI Y., LONG J., YU T. *et al.*, "A hybrid BCI system for 2-D asynchronous cursor control", *Engineering in Medicine and Biology Society, Annual International Conference of the IEEE*, pp. 4205–4208, 2010.

[LOT 08] LOTTE F., Study of electroencephalographic signal processing and classification techniques towards the use of Brain–Computer Interfaces in virtual reality applications, PhD thesis, INSA Rennes, 2008.

[LOT 10] LOTTE F., VAN LANGHENHOVE A., LAMARCHE F. *et al.*, "Exploring large virtual environments by thoughts using a Brain–Computer Interface based on motor imagery and high-level commands", *Presence: Teleoperators and Virtual Environments*, vol. 19, pp. 54–70, 2010.

[MAC 92] MACKENZIE I.S., BUXTON W., "Extending fitts' law to two-dimensional tasks", *Proceedings of the SIGCHI Conference on Human Factors in Computing Systems*, pp. 219–226, 1992.

[MCF 08] MCFARLAND D., KRUSIENSKI D., SARNACKI W. *et al.*, "Emulation of computer mouse control with a noninvasive brain–computer interface", *Journal of Neural Engineering*, vol. 5, pp. 101–110, 2008.

[MER 13] MERCIER-GANADY J., LOUP-ESCANDE E., GEORGE L. *et al.*, "Can we use a Brain–Computer Interface and manipulate a mouse at the same time?", *Proceedings of the ACM Symposium on Virtual Reality Software and Technology*, pp. 69–72, 2013.

[NIJ 09] NIJHOLT A., BOS D.P.-O., REUDERINK B., "Turning shortcomings into challenges: Brain–Computer Interfaces for games", *Entertainment Computing*, vol. 1, no. 2, pp. 85–94, 2009.

[NIJ 13] NIJHOLT A., GÜRKÖK H., "Multi-brain games: cooperation and competition", STEFANIDIS C., ANTONA M. (eds), *Universal Access in Human-Computer Interaction. Design Methods, Tools, and Interaction Techniques for eInclusion*, Springer, pp. 652–661, 2013.

[NOR 86] NORMAN D.A., DRAPER S.W., (eds), *User Centered System Design: New Perspectives on Human-Computer Interaction*, Lawrence Erlbaum, 1986.

[PFU 10] PFURTSCHELLER G., ALLISON B.Z., BAUERNFEIND G. *et al.*, "The hybrid BCI", *Frontiers in Neuroscience*, vol. 4, no. 3, 2010.

[PHI 07] PHILIPS J., DEL R.MILLAN J., VANACKER G. *et al.*, "Adaptive shared control of a brain-actuated simulated wheelchair", *IEEE International Conference on Rehabilitation Robotics*, pp. 408–414, 2007.

[REB 07] REBSAMEN B., TEO C.L., ZENG Q. *et al.*, "Controlling a wheelchair indoors using thought", *Intelligent Systems*, vol. 22, pp. 18–24, 2007.

[RIC 03] RICHARDSON J.C., SWAN K., "Examining social presence in online courses in relation to students' perceived learning and satisfaction", *Journal of Asynchronous Learning Networks*, vol. 7, pp. 68–88, 2003.

[SCH 07a] SCHERER R., MÜLLER-PUTZ G.R., PFURTSCHELLER G., "Self-initiation of EEG-based brain–computer communication using the heart rate response", *Journal of Neural Engineering*, vol. 4, pp. 23–29, 2007.

[SCH 07b] SCHLÖGL A., KEINRATH C., ZIMMERMANN D. *et al.*, "A fully automated correction method of EOG artifacts in EEG recordings", *Clinical Neurophysiology*, vol. 118, no. 1, pp. 98–104, 2007.

[VAU 06] VAUGHAN T., MCFARLAND D., SCHALK G. *et al.*, "The wadsworth BCI research and development program: at home with BCI", *IEEE Transactions on Neural Systems and Rehabilitation Engineering*, vol. 14, pp. 229–233, 2006.

[WAN 08] WANG Y., GAO X., HONG B. *et al.*, "Brain–Computer Interfaces based on visual evoked potentials", *Engineering in Medicine and Biology Magazine*, vol. 27, pp. 64–71, 2008.

[ZAN 10] ZANDER T.O., GAERTNER M., KOTHE C. *et al.*, "Combining eye gaze input with a Brain–Computer Interface for touchless human-computer interaction", *International Journal of Human-Computer Interaction*, vol. 27, no. 1, pp. 38–51, 2010.

[ZAN 11] ZANDER T.O., KOTHE C., "Towards passive brain–computer interfaces: applying brain–computer interface technology to human–machine systems in general", *Journal of Neural Engineering*, vol. 8, no. 2:025005, 2011.

13

Brain Training with Neurofeedback

13.1. Introduction

Neurofeedback (NF) is a *biofeedback* technique that involves providing information to an individual about his or her brain activity in the form of visual, auditory or tactile feedback, updated in real time. For example, individuals can be shown a gauge showing their frontal beta activity, as an indicator of their concentration level. NF allows subjects to exploit their *self-regulation* ability to develop an optimal mental strategy for achieving an objective expressed in terms of brain activity. The sensory feedback informs the user whether cerebral activity is getting closer or further away from the objective. Thus, NF is a way of performing *operant conditioning* of brain activity, with feedback acting as reinforcement. The goal of NF training for users is to learn to improve how they regulate certain aspects of their cerebral activity, so that this activity may be reorganized sustainably. NF is composed of diferent training protocols, each targeting specific brain patterns, such as those related to concentration, relaxation, the imagination of movements or the visualization of positive memories. It may be used as a tool for exercising the brain and optimizing performance in healthy subjects, or for brain *rehabilitation* in the case of patients with brain disorders. Over the 50 years of its existence, NF has been studied in a very large range of clinical and non-clinical applications (see section 13.5). Even so, its effectiveness in each

Chapter written by Lorraine PERRONNET, Anatole LÉCUYER, Fabien LOTTE, Maureen CLERC and Christian BARILLOT.

of these applications remains to be proven and properly differentiated from a placebo effect. Currently, the most convincing application of NF is in the treatment of attention-deficit hyperactivity disorder (ADHD) in children, for which randomized controlled trials (RCT) and an initial set of meta-analyses have been published [ARN 09]. A number of studies also seem to agree on the potential of NF for treating pharmaco-resistant epilepsy [TAN 09]. Despite the preliminary nature of this research, some practitioners in the United States and Canada already use NF, claiming to be able to treat a vast panel of physical, mental and cognitive disorders. Thus, NF exists in two different forms: the current state of research and its practical applications. These two forms are not necessarily mutually representative. In France, the clinical research community has long been suspicious of this technique, and has only recently started to consider it more closely. Let us note that at this point the neuroptimal method, originating in Canada and claiming to be based on NF, has started to gain momentum in France over the last few years. Our position on this method is that it is not an NF technique, as it claims to reorganize the brain passively, whereas NF requires the subject to undertake a conscious and voluntary learning process. Furthermore, the method is not supported by any officially recognized study. Although other biofeedback techniques are based on the activity of the peripheral nervous system (cardiac rhythm, muscle tension and skin conductivity), NF focuses on the activity of the central nervous system (CNS). This activity can be measured in real time using various non-invasive methods, such as electroencephalography (EEG), functional magnetic resonance imaging (fMRI), near-infrared spectroscopy (NIRS) and magnetoencephalography (MEG). Most of the work published since the early days of NF in the 1960s has been based on EEG. There is a considerable body of literature in this area; we shall denote it EEG-NF, but it is sometimes also referred to as "EEG biofeedback". Although EEG is currently the only modality used by NF practitioners, it is limited by a lack of specificity due to its low spatial resolution. Research has therefore turned to other modalities that allow the activity of different regions of the brain to be more precisely targeted. Since the turn of the millennium, the strongly dynamic research into MRI-NF appears to hold promising results for treating depression [YOU 14b] and chronic pain [DEC 05] by virtue of its capacity to provide real-time imagery of activity in deep brain structures with high spatial resolution. More recently, it was demonstrated that NIRS-NF [MIH 12] and MEG-NF [SUD 11] are technologically feasible.

Like BCIs, NF relies on a closed loop that exploits brain activity in real time, specifically by acquisition of a signal originating from the brain, signal preprocessing (noise removal, filtering), extraction of relevant features that allow the state or intent of the subject to be recognized and translation into feedback to close the loop and to allow the subject to adapt in real time. Even though the two approaches of NF and BCIs share very similar technologies, their original purposes are very different: BCIs allow the subject to control an *external object* such as a computer, an orthosis or a robotic limb, whereas NF allows subjects to acquire control over *themselves*. Although some BCIs, e.g. spontaneous BCIs, involve a learning process, and so require the subject to perform cerebral self-regulation, this self-regulation is ultimately not the purpose of the exercise. Both NF and BCIs emerged during the sixties and seventies from works studying operant conditioning of brain activity, revealing that humans [KAM 62, STE 72] and animals [WYR 68, FET 69] are capable of learning to generate specific brain patterns, assuming that they are informed or rewarded when they are successful in doing so. This reward is indispensable for animals [FET 69] and secondary but recommended for human subjects. Following these discoveries, two multidisciplinary communities formed, uniting clinicians, neurologists, psychologists and engineers among others, each with the objective of meeting a specific set of needs:

– the objectives of the NF community were the self-regulation of brain patterns for treating neurological and psychiatric disorders and the optimization of performance. The NF community diversified and refined the targeted brain patterns, demonstrated the benefits of learning and also studied self-regulation processes and the optimization of psychoexperimental learning factors;

– the BCI community instead focused on implementing control and communication interfaces for handicapped persons, investing their efforts into developing tools for signal processing and for classifying brain patterns to optimize reliability and degrees of freedom of BCIs.

Over the past few years, invasive BCIs have attracted attention after several spectacular demonstrations with tetraplegic patients [PFU 00] [HOC 12], and non-invasive BCIs have become increasingly popular because of new applications in video games and virtual reality [LÉC 08]. On the other hand, NF still suffers from the reputation of new age technique in some scientific circles, a black mark inherited from the outcry of the media in early days of NF in the 1960s (see section 13.3) that is proving difficult to overcome. As a field of study, NF can appear impenetrable due to the abundance of unreliable information and the diversity of existing protocols and applications. There has, nonetheless, been a revival in the number of publications in recent years, and a renewed approach indicating that interest in the potential of NF is picking up.

Section 13.2 below describes the design of a (practical) NF training program, and the typical course of a (practical and research-oriented) NF session, as well as the learning mechanisms underlying NF. Section 13.3 retraces the history of NF, explaining the origin of its questionable reputation and providing a foothold for understanding the diversity of existing approaches. Section 13.4 discusses how the fields of NF and BCIs might potentially overlap in future with the development of "restorative" BCIs. Finally, section 13.5 presents the range of applications of NF and summarizes the state of research of some of its major clinical applications.

13.2. How does it work?

13.2.1. *Design of an NF training program*

NF is composed of protocols that target different brain patterns assumed to underpin specific sets of physical, mental or cognitive functions. We shall describe NF such as it is practiced in NF clinics, whether or not the scope is clinical in nature. Currently, only EEG-based variants of NF are used in practical applications. NF training programs are organized into several stages, the practical aspects of which vary depending on the needs of the subject and the equipment and software used by the practitioner. These stages are generally as follows:

1) *Diagnosis*: this first stage involves identifying the symptoms suffered by the subject, or the function that the subject wishes to improve, in order to establish the target of the training process in functional terms. Generally, this begins with an examination of the patient's medical history, followed by, depending on the target disorder or the desired objective, a series of specific, standardized tests administered in the form of questionnaires or psychological, neurological or physiological examinations.

2) *Choice of NF protocol*: in the second stage, an EEG-NF protocol is defined to suit the subject's needs; in other words, the electrical activity in

the brain that the training process will target is established. This activity corresponds to a certain brain measurement, which is assumed to be indicative of or to act upon the function that the subject wishes to improve. During training, this measurement will be displayed to the subject in the form of feedback; the subject will then attempt to manipulate the measurement toward a desired value or in a desired direction. The simplest type of measurement available is the power in a frequency band measured at an electrode, for example the power of mu waves in the motor regions for a task of movement imagination. However, more complex metrics that reflect the exchange of information between two or more regions may also be used, such as the coherence or asymmetry measured between two electrodes. Over more than 50 years of research, a large number of protocols have been studied, and are now offered by NF clinics in response to a wide array of conditions (although official research has not yet provided sufficient proof of their effectiveness). The definition of the protocol can be made using *a priori* knowledge from the literature and the experience of the practitioner, or by a *personalized* process for each patient, or a combination of these two approaches. The personalized approach requires additional tests, and particular tools and expertise. For example, certain practitioners use quantified EEGs, which involves comparing the subject's EEG to a database of healthy subjects to detect potential irregularities which can then be targeted during training [BUD 09].

3) *Session planning*: there must be sufficiently many sessions for the first effects to be visible (the subject explores the strategies), and also so that the changes endure in the long term (the subject repeats and attempts to maintain an effective strategy; with practice this becomes ever easier). The required number of sessions (between 20 and 40 [HAM 11] [GRU 14c]) depends on the severity of the subject's symptoms, his or her learning skills and motivation levels, but also on the chosen target pattern, which may be more or less difficult to master.

13.2.2. *Course of an NF session: where the eyes "look" at the brain*

Once the NF training program has been established, the training sessions can commence. In this section, we shall describe the typical course of an NF session, which holds both in practice and in research contexts. Figure 13.1 shows an example of an EEG-NF environment. Over the course of a session of NF, which generally lasts less than 1 h to account for the subject's fatigue levels, the subject sits in front of a screen (except for fMRI-NF where the subject is lying down in the MR tube), potentially with earphones or, in rarer occasions, connected to a tactile interface. The subject is asked to perform a mental task, such as concentrating, relaxing, thinking of something positive or imagining a movement in the right hand. Sometimes, the subject is not given any explicit instruction [KOB 13a] other than to attempt to control the feedback using the mental strategy that he or she finds most effective. The equipment (traditionally EEG, but alternatively fMRI, NIRS, MEG) records brain signals, which are then processed by an algorithm. The algorithm cleans the signals of any artifacts created by movement or noise from the environment. It then extracts a relevant metric related to the requested task or targeted function, and translates this metric into feedback that indicates to the subject how well the task was performed (as a score), and in which direction he or she must focus subsequent efforts (like an arrow). The feedback acts as an objective indicator of what is happening in subject's mind. Typically, the screen displays a gauge that fills or empties depending on the subject's performance, but various other forms of feedback are also possible such as sounds with different pitches, the motion of an orthesis or an interface with a video game or virtual reality. The practitioner remains at the subject's side, observing and guiding the subject by providing advice about mental strategies, encouragement or by helping the subject to relax. The practitioner also monitors the subject's performance, and if necessary adjusts the parameters of the protocol, such as the difficulty of the task (reinforcement threshold) or the properties of the feedback. In order to ensure that skills acquired during training *transfer* to everyday life, the subject may be asked to attempt to regulate activity without feedback at certain points in the session, and also to practice outside of the sessions.

13.2.3. *A learning procedure that we still do not fully understand*

Even if NF was shown to have positive effects on behavior and cognition more than 50 years ago [KAM 62], the mechanisms underlying brain self-regulation during NF training are barely understood [BIR 13, NIV 13]. This lack of understanding at a fundamental level is one of the reasons why it has been difficult to establish good practices. Methodological aspects, even essential ones such as the optimal form and frequency of feedback, are still being debated. NF is generally viewed as a *learning procedure* based on principles of *operant conditioning* [SHE 11] (the feedback acts as reinforcement) and *neuroplasticity* (via the training program and session repetition) and relying on a *voluntary and conscious* involvement on the part of the subject. But these principles do not seem to fully describe the NF process. Recently, certain authors have questioned the central role of operant conditioning and conscious and voluntary action of the subject within the NF procedure. They suggest that NF is instead an instance of *implicit* learning (automatic subconscious process, similar to acquiring the ability to perform a new movement) [BIR 13, NIN 13], or even a *dual-process* mechanism with both conscious and subconscious components [STR 14].

Figure 13.1. *Example of an EEG-NF environment*

A small set of studies have begun to explore the brain mechanisms that are at play during NF session, as well as the factors that might affect the subject's performance. For example, the study performed by Ninaus [NIN 13] showed that when individuals attempt to control a gauge that they believe is an indicator of their own brain activity, they activate regions of the brains that are involved in *self-referential* processes and *cognitive control*, which does not occur when the subject observes the gauge passively. Another study

[KOB 13a] observed that individuals without any specific mental strategy could increase their sensorimotor rhythm (SMR) more successfully than those who reported adhering to a specific strategy, an observation that supports the hypothesis of implicit learning. Questioning the influence of *mental strategies* and *explicit instructions* appears to be particularly relevant when considering the component of voluntary involvement of the subject in the NF process. Other factors non-specific to NF can also influence the subject's performance: motivation levels, concentration levels, moods, relaxation, the subject's ability to limit muscle-related artifacts, the subject's loci of control or even the therapist–patient relationship. Understanding the role of these specific and non-specific factors in facilitating NF learning should allow more effective NF protocols to be established, and better-controlled studies to be designed [STR 14].

13.3. Fifty years of history

13.3.1. *A premature infatuation*

The principle of NF was demonstrated for the first time in humans by James Kamiya at the University of Chicago in 1962 [KAM 62]. In order to discover whether individuals are capable of recognizing spikes of occipital alpha activity in their own brains, he attempted to train a volunteer by providing verbal confirmation each time that the volunteer produced alpha activity. As a result, the subject not only became capable of recognizing the apparition of alpha waves, but also acquired the ability to generate them at will. At the time, research into modified states of consciousness was particularly fashionable. One important study reported that zen monks had particularly elevated alpha activity levels during meditation [KAS 66]. Working from this observation, alpha wave NF was further explored with the goals of reducing anxiety [HAR 78] and stress, as well as inducing states of relaxation or deep meditation and stimulating creativity. In parallel to this research into alpha NF, in 1969, Barry Sterman from the of University California discovered by accident the therapeutic potential of NF for treating epilepsy [STE 69]. During a study that he was performing for NASA on the epileptogenic properties of a certain type of rocket fuel (hydrazine), he observed that among 50 cats exposed to hydrazine, 10 were particularly resistant to seizures. It just so happened that these same cats had been trained to increase their (SMR, 11–15 Hz) by NF in a previous series of experiments

[WYR 68]. The fact that this result was observed in animals is particularly significant, as it proves that the effects of NF training cannot be simply reduced to a placebo effect. Encouraged by this surprising discovery, Sterman extended the study of the protocol to humans [STE 72]. For 3 months, twice a week, he trained one of his female colleagues, 23 years, who was suffering from generalized epileptic seizures. After training was completed, he observed a reduction of seizures coupled with an increase in SMR rhythm and a reduction in slow waves. The treatment was continued until the subject recovered completely, and the young woman even managed to obtain her driving license. In 1976, Lubar [LUB 76] showed the benefits of the SMR protocol with symptoms of hyperactivity and distractability in a child with hyperkinetic disorder. He started by training the child to increase his SMR rhythm, and observed an improvement in the symptoms. Then, conversely, he trained the child to reduce his SMR rhythm using the experimental ABA model, and the symptoms resurfaced. He then successfully replicated this study with four children suffering from ADHD [SHO 79].

The results of Kamiya's research were published in 1968 in an article of *Psychology Today* [KAM 68], and the idea that alpha NF could be used to attain a meditative state aroused a great deal of enthusiasm. Unfortunately, this publication, while introducing the concept of NF to the general public, also triggered a premature and uncontrolled propagation of the technology. Even though NF was in a stage of early infancy, an industry rapidly formed around it, producing NF kits that promised to allow users to learn to control their brain waves, and reach illumination without needing to invest years into the practice of meditation. However, these poor-quality devices were essentially smokescreens, and the validity of the link between NF alphas and meditation had not been properly established by scientific studies.

13.3.2. *Diversification of approaches*

Thus, the beginnings of NF were marked by scientific discoveries that were both surprising and promising, but also by the parallel development of a new age industry-based without rigorous foundations, which caused NF to be marginalized and relegated to the status of pseudoscience. The resulting lack of finance support placed a considerable brake upon research, which was confined to a small set of laboratories working in isolation. Despite the poor reputation afflicting NF within the scientific community at the end of the seventies, research continued, initially in sparse increments, before gradually expanding into a dynamic field of research with an ever-growing number of publications. The available protocols and practical procedures diversified, benefiting from technical advancements in the quality of EEG devices, brain imaging (fMRI), computer processing capacities and scientific advancements in neurophysiology and electrophysiology. Protocols were developed to extend the range of targets to the rest of the EEG frequency spectrum (other than alpha and SMR), including the outer frequency bands, which generally require specialized measuring equipment. The hypnagogic state induced by the alpha/theta protocol is thought to have potential for treating depression and anxiety resulting from alcoholism, and posttraumatic stress disorder [PEN 89], as well as for stimulating creativity [GRU 14b]. NF with high-frequency gamma waves, which has only been studied recently as it requires high-performance measuring equipment, seems on the other hand to act upon the cognitive performance [KEI 10]. In the 1980s, a new type of NF developed, breaking from traditional forms of NF that use EEG frequencies, based on the studies on self-regulation of slow cortical potentials (SCP) [ELB 12]. These potentials are well known as indicators of the level of cortical excitability. In the 1990s, NF of SCP was studied in populations of patients suffering from ADHD [ROC 90] and epilepsy [ROC 93]. Despite promising results, the use of the technique was limited for the longest time for reasons of equipment and the level of mastery required to correctly measure SCP [STR 09]. In the early 1990s, certain practitioners pursuing the idea of normalizing the EEG activity of their patients began to use quantitative EEG to choose the target of NF [BUD 09]. With this development, the targets of the protocols were extended to include metrics other than the amplitude of a frequency band in a given region, traditionally measured with a monopolar setup. Certain practitioners used bipolar setups, for example to target the coherence between two regions if it is identified as atypical by quantitative EEG [BUD 09]. Toward the end of the nineties, Thatcher [THA 98] introduced the Z-score (EEF-)NF (BrainMaster) based on data obtained from quantitative EEG, which allows multiple different metrics to be trained simultaneously, such as the absolute power, the power ratio, the coherency, the phase delay or even the asymmetry. Breaking away from the classical approach of EEG-NF that is based on using one or two EEG channels, the LORETA (EEG-)NF or tomographic (EEG)-NF suggested by Congedo in 2004 [CON 04] uses a fully equipped EEG headset with 19 electrodes for reconstructing and targeting deep sources of activity in real time with the

LORETA method. The behavioral, cognitive and electrophysiological effects of this type of NF were first observed and described by Cannon [CAN 09]. Later, Liechti [LIE 12] evaluated the effectiveness of tomographic NF of the anterior cingulate cortex for treating ADHD, and Maurizio *et al.* [MAU 14] compared it with biofeedback of electromyographic activity. The Z-score (EEG-)NF was recently combined with tomographic NF to give the Z-score LORETA (EEG-)NF [KOB 13b, THA 13].

Finally, in the past few years, the NF community has begun to explore new modalities. The development of real-time fMRI [COX 95] led to the noteworthy milestone of fMRI-NF [YOO 02]. The major benefit of this technique over EEG-NF is that it provides high spatial resolution access to deep brain structures housing complex functions such as emotions, memory and pain. fMRI-NF can target any brain region with millimeter precision, allowing users to train their BOLD level, reflecting oxygen consumption, which is indirectly correlated with neuronal activity. The feasibility of this kind of NF was first shown in the anterior cingulate cortex [WEI 03], and since then has been demonstrated in numerous other regions of the brain [SUL 13]. In 2005, deCharms [DEC 05] reported findings of pain reduction in patients with chronic pain after learning to control BOLD activity in the rostral anterior cingulate cortex over the course of four sessions. Recent work has suggested promising results of fMRI-NF in the amygdala for depression [YOU 14b]. fMRI-NF is currently only used in the context of research, and requires wider scale study before it can be integrated into therapeutic programs. More recently, the exploration of SPIR-NF [MIH 12] has begun, which could potentially provide a portable alternative method that is less expensive than fMRI-NF, although restricted to superficial brain structures and with lower spatial resolution. Finally, MEG-NF [SUD 11] is still in its infancy, and wide-scale application is not currently conceivable due to the costs involved and the low number of MEG equipment sets in operation (only five in France). MEG could, however, be used as a preliminary to EEG-NF in order to define more precisely the target of treatment. Figure 13.2 summarizes the emergence of the various types of NF, and the therapeutic trials with which they are canonically associated.

13.4. Where NF meets BCI

In the scientific literature, the various ways that the terms NF and BCI are used reveals the diversity of authors' conceptions about these topics according to their original fields of study. For example, sometimes BCI is used to refer to the technology and NF is used to refer to the application [YAN 08], which could be interpreted to imply that BCIs are an implementation of the concept of NF. NF is also presented relatively often as a special case of BCIs [WOL 11] with a minimal closed loop (no command sent to an external object). In some instances, albeit less frequently, this picture is inverted, so that BCIs are presented as a special case of NF [NEU 10, GRO 11]. Indeed, the behavior of objects in BCIs could itself be viewed as a kind of feedback. Finally, in some cases both terms are used in tandem, such as in some reviews of the state of the art that group the therapeutic applications of NF and BCIs into the same category [BIR 09]. This particular usage of terminology rightly highlights the similarities of the two approaches and the fact that clinical applications of BCIs have historically been derived from work on NF.

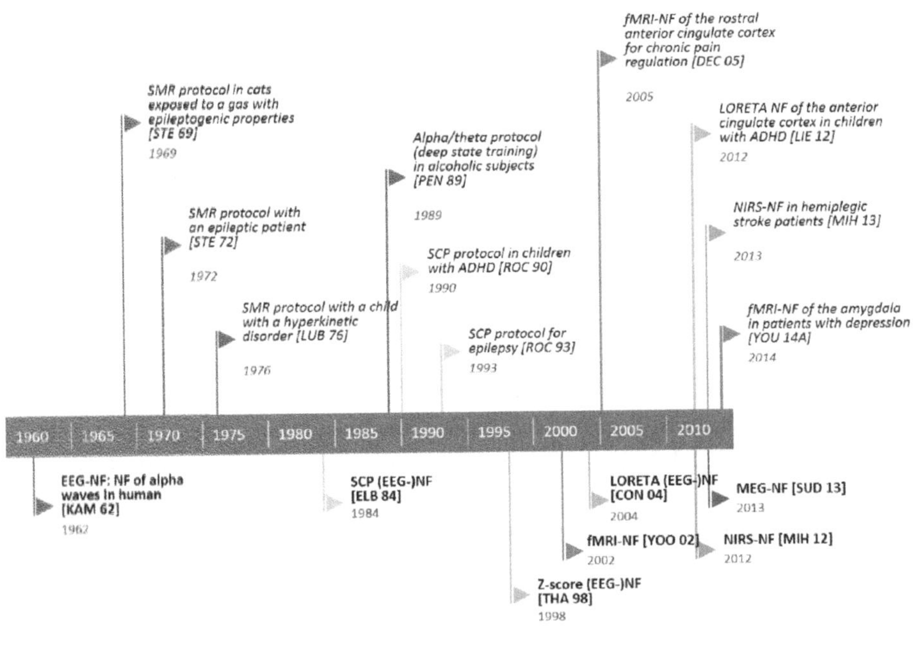

Figure 13.2. *Fifty years of NF history; below: the proofs of feasibility of various types of NF; above: their canonical therapeutic trials. For a color version of the figure, see www.iste.co.uk/clerc/interfaces1.zip*

Recently, the definition of BCI has been extended to account for new types of applications [ALL 11]. The modern definition of a BCI was given by

Wolpaw in 2011 [WOL 11]: "*A BCI is a system that measures CNS activity and converts it into artificial output that replaces,* restores, *enhances, supplements or* improves *natural CNS output and thereby changes the ongoing interactions between the CNS and its external or internal environment.*" Unlike the traditional definition of BCI [WOL 02], this definition admits the family of applications linked to brain rehabilitation, which until recently had fallen exclusively under NF. This definition of BCIs therefore fully encompasses NF. The specific term "restorative BCI" has even emerged to describe BCIs designed for purposes of brain rehabilitation, as opposed to the term "assistive BCI", describing BCIs that serve purposes of communication and control. These "restorative BCIs" are thus equivalent to the definition of NF. The reason that this term was introduced despite the prior existence of the term "neurofeedback" is likely that the BCI community has chosen to boycott the label "neurofeedback" due to its connotation of pseudoscience. In BCI literature, restorative BCIs currently focus mostly on the motor rehabilitation of stroke victims [BIR 09, GRO 11, ANG 15, DOK 15]. Note that patients with motor deficits (cerebral palsy, strokes, amyotrophic lateral sclerosis, multiple sclerosis, muscular dystrophy, bone marrow lesions, etc.) were historically the primary target audience of the first (assistive) BCIs. But applications of restorative BCIs also tend to involve restoring functions that are physical (other than motor), cognitive and emotional in nature [WOL 11], just like NF.

13.5. Applications

The traditional purpose of NF has always had two components; on the one hand, oriented toward *therapeutic* applications for treating *psychiatric and neurological disorders*, and on the other hand, toward *optimizing the performance* of healthy subjects. Since the brain is the control center of physical, mental and cognitive functions, and NF acts directly upon brain mechanisms that govern or contribute to these control processes, the potential scope of NF is vast. This is reflected in the panel of applications for which NF has already been explored:

– therapeutic applications: ADHD, epilepsy, depression, anxiety, learning disorders, sleeping disorders, autism, posttraumatic stress disorder, addiction, chronic pain, tinnitus, migraines, Alzheimer's, Parkinson's, schizophrenia, etc. [BIR 09, HAM 11, NIV 13, WYC 14];

– non-therapeutic applications: mental rotation, attention, memory, visual skills, musical performance, performance of surgical tasks, sportive performance, cognitive deterioration and meditation [GRU 14a, GRU 14b, SCH 15].

Despite the number and variety of NF studies, the results obtained are preliminary in nature, except in the domain of ADHD, where NF is currently in the process of achieving sufficient levels of proof. Indeed, most NF studies are case studies, or have methodological flaws that limit the scope of their conclusions. Maintaining quality levels in NF studies aiming to demonstrate the effectiveness of an NF protocol in improving a given disorder or function is therefore a substantial multidisciplinary *methodological challenge*. NF has long suffered from a lack of scientific credibility inherited from the rapid failure of alpha NF in the 1970s, but also from the poor quality of early studies that were not reproducible, with very few subjects in poorly controlled experimental conditions. Today, it has been clearly established that in order for an NF protocol to be proven effective in a given application, it must first be subjected to rigorous experimentation that observe the principles of evidence-based medicine: significant numbers of subjects, use of control groups (in particular placebo groups [ARN 14]), randomization, double-blind trials, studies by multiple research groups, monitoring of physiological, psychological and behavioral effects in the short, medium and long term. Given that NF training programs often require over 20 sessions to obtain sustainable results, and that these sessions must be multiplied by the number of subjects in the study, the resources required to properly meet all of these requirements soon become colossal. The only domain in which these requirements have started to be met is that of ADHD.

By exploiting the self-regulation capacity of its users, NF could potentially find its place as a non-invasive or *complementary* alternative to other treatments, such as pharmacological treatments, neurosurgery, psychotherapy and passive stimulation techniques (such as transcranial magnetic stimulation (TMS) and transcranial direct current stimulation). Below, we shall describe the state of the art for the major therapeutic applications of NF.

ADHD: ADHD is the clinical application of NF that has been studied in the most depth till now. The protocols that have been shown to be beneficial are the theta/beta (inhibition of theta and facilitation of beta at the fronto-central electrodes), SMR and SCP protocols. One meta-analysis

[ARN 09] of over 15 RCT studies with a total of 1,194 patients concluded that EEG-NF could be considered an "effective and specific" alternative to usual treatments based on stimulants (methylphenidate and amphetamines), which are limited by their short-term mode of action and potential side effects. The authors of the meta-analysis reported a large effect size on symptoms of inattention and impulsiveness, and a medium effect size on hyperactivity. However, the effect sizes were found to be smaller in better-controlled studies, i.e. studies with a placebo control group [ARN 09]. Additional RCTs with placebo groups are still necessary, particularly double-blind studies, to more precisely identify the effectiveness of NF. As for the duration of the treatment effect, several studies showed that the effects of the SCP and theta/beta protocols persisted after the last session of NF, for 6 months [GAN 09, LEI 07] and for 2 years [GAN 09].

Epilepsy: roughly one-third of patients suffering from epilepsy do not respond to anticonvulsants and cannot be treated by operation due to the localization of their epileptic foci. NF could represent a viable alternative for these patients. Epilepsy was the first therapeutic application of NF to be discovered [STE 69]. The protocols that seem best adapted to epileptic patients are the SMR and SCP protocols. One meta-analysis [TAN 09] of 174 patients over 10 studies (nine SMR and one SCP) showed that the frequency of seizures decreased in 74% of the patients. One study even observed persistent effects 9 years after treatment with NF [STR 14]. Despite these encouraging results, larger scale controlled trials are required to prove that NF is effective in cases of pharmacoresistant epilepsy.

Depression: unlike existing therapies (psychotherapy, antidepressants, electroshock therapy) and more modern alternatives (vagus nerve stimulation, TMS), one of the unique aspects of NF is that patients suffering from depression are placed in the role of actors in their healing process, demonstrating to themselves that they have the capacity to influence their own brain activity, and consequently their psychological states. The EEG-NF protocol that has been studied in most depth for depression aims to regulate the frontal asymmetry of alpha waves based on the hypothesis that in some patients suffering from depression the right prefrontal lobe associated with withdrawal behavior is hypoactive, whereas the left prefrontal lobe associated with approach behavior is hyperactive. Although several studies appear to report an improvement in depression scores following the application of this protocol [CHO 11, PEE 14], these studies have a number of methodological flaws such as low patient numbers, low initial depression scores, limited number of evaluation criteria for describing effects on depression and lack of control for unspecific effects. Furthermore, the specificity of the frontal asymmetry marker is still subject to debate. As emotional circuits involve deep and complex networks, it is possible that traditional EEG-NF is not sufficient to regulate this kind of circuit. One LORETA-NF study [PAQ 09] showed that reducing rapid beta wave activity (18–30 Hz) in the corticolimbic/paralimbic regions of patients suffering from major depression was correlated with an improvement in depression-related symptoms. fMRI-NF might also be promising for treating depression, giving its capacity to target the deep regions of the emotional circuit such as the amygdala and the ventrolateral prefrontal cortex [YOU 14b]. A recent study also showed that it is possible to simultaneously combine EEG-NF of frontal asymmetry and fMRI-NF of the amygdala [ZOT 14], which could allow the effects of both protocols to be combined.

Motor rehabilitation following a stroke: It is an established fact that the imagination of a movement activates regions and circuits similar to those activated by effectively performing the same movement [SHA 13]. By practicing imagining movements, individuals with paralysis after a stroke directly utilize the damaged motor regions in the brain, stimulating neuronal plasticity. It is assumed that practice can help them progressively recover motor function. To allow patients to overcome the phenomenon known as "acquired non-use' and to achieve actual recovery of motor control in the affected limb, closing the sensorimotor loop during NF seems crucial [DOK 15]. Visual or audio feedback is generally not sufficient for this, and instead tactile feedback should be used. This feedback should be well calibrated to fit the subject's intended movement, for example using an orthesis [SHI 11, RAM 13] or by combining NF with function electrical stimulation [YOU 14a].

NF for controlling BCIs: as remarked by [LOT 13], "using BCIs is a skill". Indeed, before being able to operate a BCI (especially a spontaneous BCI) with a given brain pattern, e.g. by imagining moving the left or right hand, subjects must first learn to generate this pattern reliably [HWA 09]. Subjects can achieve this through hours of mental practice, but it is difficult for them to be certain that they are practicing the correct task, and that they are orienting their efforts in the correct direction. NF provides users with an objective representation of the pattern, allowing them to practice controlling it in more favorable learning conditions. NF-based practice is therefore an indispensable preliminary stage for the usage of certain BCIs, such as spontaneous BCIs.

13.6. Conclusions

NF is a technique that involves measuring the brain activity of a subject and communicating this information to the subject in real time so that he or she may learn to better control one specific aspect of it. Until recently, the primary distinction between this technique and BCIs was essentially the fact that the objective of NF is self-control, whereas BCIs (in the classical sense of assistive BCIs) aim to control objects external to the user. Today, this distinction has begun to fade, as the field of applications of BCIs has been extended to include BCIs for brain rehabilitation, given the name of "restorative BCIs". We might therefore hope to see the NF and BCI communities come closer together in future, thanks to the transfer of experience and knowledge. For example, the NF community might benefit from the more sophisticated processing tools and feedback metaphors developed for BCIs, while the BCI community might benefit from the learning and brain self-regulation principles used in NF.

Although certain forms of EEG-NF are already used by practitioners who boast of innumerable benefits, NF is still an experimental technique. Among all of the EEG-NF protocols suggested over the course of 50 years in response to a very wide panel of applications, only the SCP, SMR and theta/beta protocols for treating ADHD have begun to attain sufficient levels of proof. Outside of this special case, it seems that traditional NF approaches based on EEG with one or two channels are limited by the quality and the low specificity of the recorded information. Thus, in order to better target the activity of certain regions of the brain, researchers have developed new NF techniques that use larger numbers of EEG electrodes (LORETA-NF with a full EEG headset) or rely on other equipment (fMRI-NF, MEG-NF, SPIR-NF). The most noteworthy of these newer techniques is fMRI-NF, which allows activity in deep brain regions to be measured with high spatial resolution. fMRI-NF already shows promise for treating emotional disorders [YOU 14b], chronic pain [DEC 05] and other disorders [SUL 13]. Studies questioning the mechanisms at work in NF are another active area in the current state of NF, which should lead to the development of more effective protocols. Lastly, in order to prove the effectiveness of these protocols in specific conditions and to justify their integration into treatment programs, large-scale RCT will be indispensable. There are currently far too few such studies.

13.7. Bibliography

[ALL 11] ALLISON B., "Trends in BCI research", *XRDS: Crossroads, The ACM Magazine for Students*, vol. 18, no. 1, pp. 18–22, 2011.

[ANG 15] ANG K.K., GUAN C., "Brain–computer interface for neurorehabilitation of upper limb after stroke", *Proceedings of the IEEE*, vol. 103, no. 6, pp. 1–10, 2015.

[ARN 09] ARNS M., DE RIDDER S., STREHL U., "Efficacy of neurofeedback treatment in ADHD: the effects on inattention, impulsivity and hyperactivity: a meta-analysis", *Clinical EEG and Neuroscience*, vol. 40, no. 3, pp. 180–189, 2009.

[ARN 14] ARNS M., HEINRICH H., STREHL U., "Evaluation of neurofeedback in ADHD: the long and winding road", *Biological Psychology*, vol. 95, no. 1, pp. 108–115, 2014.

[BIR 09] BIRBAUMER N., RAMOS MURGUIALDAY A., WEBER C. *et al.*, "Neurofeedback and brain–computer interface clinical applications," *International Review of Neurobiology*, vol. 86, pp. 107–117, 2009.

[BIR 13] BIRBAUMER N., RUIZ S., SITARAM R., "Learned regulation of brain metabolism", *Trends in Cognitive Sciences*, vol. 17, no. 6, pp. 295–302, 2013.

[BUD 09] BUDZYNSKI T., BUDZYNSKI H., EVANS J. *et al.*, *Introduction to Quantitative EEG and Neurofeedback: Advanced Theory and Applications*, Elsevier Science, 2009.

[CAN 09] CANNON R., CONGEDO M., LUBAR J. *et al.*, "Differentiating a network of executive attention: LORETA neurofeedback in anterior cingulate and dorsolateral prefrontal cortices,", *The International Journal of Neuroscience*, vol. 119, no. 3, pp. 404–441, 2009.

[CHO 11] CHOI S.W., CHI S.E., CHUNG S.Y. *et al.*, "Is alpha wave neurofeedback effective with randomized clinical trials in depression? A pilot study", *Neuropsychobiology*, vol. 63, no. 1, pp. 43–51, 2011.

[CON 04] CONGEDO M., LUBAR J.F., JOFFE D., "Low-resolution electromagnetic tomography neurofeedback", *IEEE Transactions on Neural Systems and Rehabilitation Engineering*, vol. 12, no. 4, pp. 387–397, 2004.

[COX 95] COX R.W., JESMANOWICZ A., HYDE J.S., "Real-time functional magnetic resonance imaging", *Magnetic Resonance in Medicine*, vol. 33, no. 2, pp. 230–236, 1995.

[DEC 05] DECHARMS R.C., MAEDA F., GLOVER G.H. *et al.*, "Control over brain activation and pain learned by using real-time functional MRI", *Proceedings of the National Academy of Sciences of the United States of America*, vol. 102, no. 51, pp. 18626–18631, 2005.

[DOK 15] VAN DOKKUM L.E.H., WARD T., LAFFONT I., "Brain computer interfaces for neurorehabilitation – its current status as a rehabilitation strategy post-stroke," *Annals of Physical and Rehabilitation Medicine*, vol. 58, no. 1, pp. 3–8, 2015.

[ELB 12] ELBERT T., ROCKSTROH B., LUTZENBERGER W. *et al.*, *Self-Regulation of the Brain and Behavior*, Springer, Berlin/Heidelberg, 2012.

[FET 69] FETZ E.E., "Operant conditioning of cortical unit activity", *Science*, vol. 163, no. 3870, pp. 955–958, 1969.

[GAN 09] GANI C., Long term effects after feedback of slow cortical potentials and of theta/beta-amplitudes in children with attention deficit hyperactivity disorder (ADHD), PhD Thesis, Tübingen University, 2009.

[GRO 11] GROSSE-WENTRUP M., MATTIA D., OWEISS K., "Using brain–computer interfaces to induce neural plasticity and restore function", *Journal of Neural Engineering*, vol. 8, no. 2, 2011.

[GRU 14a] GRUZELIER J.H., "EEG-neurofeedback for optimising performance. I: a review of cognitive and affective outcome in healthy participants", *Neuroscience and Biobehavioral reviews*, vol. 44, pp. 124–141, 2014.

[GRU 14b] GRUZELIER J.H., "EEG-neurofeedback for optimising performance. II: creativity, the performing arts and ecological validity", *Neuroscience and Biobehavioral Reviews*, vol. 44, pp. 142–158, 2014.

[GRU 14c] GRUZELIER J.H., "EEG-neurofeedback for optimising performance. III: A review of methodological and theoretical considerations", *Neuroscience & Biobehavioral Reviews*, vol. 44, pp. 159–182, 2014.

[HAM 11] HAMMOND D., "What is neurofeedback: an update", *Journal of Neurotherapy*, vol. 15, no. 4, pp. 305–336, 2011.

[HAR 78] HARDT J.V., KAMIYA J., "Anxiety change through electroencephalographic alpha feedback seen only in high anxiety subjects", *Science*, vol. 201, no. 4350, pp. 79–81, 1978.

[HOC 12] HOCHBERG L.R., BACHER D., JAROSIEWICZ B. *et al.*, "Reach and grasp by people with tetraplegia using a neurally controlled robotic arm", *Nature*, vol. 485, no. 7398, pp. 372–375, 2012.

[HWA 09] HWANG H.-J., KWON K., IM C.-H., "Neurofeedback-based motor imagery training for brain–computer interface (BCI)", *Journal of Neuroscience Methods*, vol. 179, no. 1, pp. 150–156, 2009.

[KAM 62] KAMIYA J., "Conditional discrimination of the EEG alpha rhythm in humans", *Meeting of the Western Psychological Association*, 1962.

[KAM 68] KAMIYA J., "Conscious control of brain waves", *Psychology Today*, 1968.

[KAS 66] KASAMATSU A., HIRAI T., "An electroencephalographic study on the zen meditation (Zazen)", *Psychiatry and Clinical Neurosciences*, vol. 20, no. 4, pp. 315–336, 1966.

[KEI 10] KEIZER A.W., VERSCHOOR M., VERMENT R.S. *et al.*, "The effect of gamma enhancing neurofeedback on the control of feature bindings and intelligence measures", *International Journal of Psychophysiology*, vol. 75, no. 1, pp. 25–32, 2010.

[KOB 13a] KOBER S., WITTE M., NINAUS M., "Learning to modulate one's own brain activity: the effect of spontaneous mental strategies", *Frontiers in Human Neuroscience*, vol. 7, 2013.

[KOB 13b] KOBERDA J.L., KOBERDA P., BIENKIEWICZ A.A. *et al.*, "Pain management using 19-electrode Z -score LORETA neurofeedback", *Journal of Neurotherapy*, vol. 17, no. 3, pp. 179–190, 2013.

[LÉC 08] LÉCUYER A., LOTTE F., REILLY R. *et al.*, "Brain–computer interfaces, virtual reality, and videogames", *Computer*, vol. 41, no. 10, pp. 66–72, 2008.

[LEI 07] LEINS U., GOTH G., HINTERBERGER T. *et al.*, "Neurofeedback for children with ADHD: a comparison of SCP and theta/beta protocols", *Applied Psychophysiology and Biofeedback*, vol. 32, no. 2, pp. 73–88, 2007.

[LIE 12] LIECHTI M.D., MAURIZIO S., HEINRICH H. *et al.*, "First clinical trial of tomographic neurofeedback in attention-deficit/hyperactivity disorder: evaluation of voluntary cortical control", *Clinical Neurophysiology*, vol. 123, no. 10, pp. 1989–2005, 2012.

[LOT 13] LOTTE F., LARRUE F., MÜHL C., "Flaws in current human training protocols for spontaneous brain–computer interfaces: lessons learned from instructional design", *Frontiers in Human Neuroscience*, vol. 7, 2013.

[LUB 76] LUBAR J.F., SHOUSE M. N., "EEG and behavioral changes in a hyperkinetic child concurrent with training of the sensorimotor rhythm (SMR) – a preliminary report", *Biofeedback and Self-Regulation*, vol. 1, no. 3, pp. 293–306, 1976.

[MAU 14] MAURIZIO S., LIECHTI M.D., HEINRICH H. *et al.*, "Comparing tomographic EEG neurofeedback and EMG biofeedback in children with attention-deficit/hyperactivity disorder", *Biological psychology*, vol. 95, pp. 31–44, 2014.

[MIH 12] MIHARA M., MIYAI I., HATTORI N. *et al.*, "Neurofeedback using real-time near-infrared spectroscopy enhances motor imagery related cortical activation", *PLoS ONE*, vol. 7, no. 3, 2012.

[NEU 10] NEUPER C., PFURTSCHELLER G., "Neurofeedback training for BCI control", *Brain–Computer Interfaces: Revolutionizing Human-Computer Interaction*, pp. 65–78, 2010.

[NIN 13] NINAUS M., KOBER S.E., WITTE M. *et al.*, "Neural substrates of cognitive control under the belief of getting neurofeedback training", *Frontiers in Human Neuroscience*, vol. 7, 2013.

[NIV 13] NIV S., "Clinical efficacy and potential mechanisms of neurofeedback", *Personality and Individual Differences*, vol. 54, no. 6, pp. 676–686, 2013.

[PAQ 09] PAQUETTE V., BEAUREGARD M., BEAULIEU-PRÉVOST D., "Effect of a psychoneurotherapy on brain electromagnetic tomography in individuals with major depressive disorder", *Psychiatry Research*, vol. 174, no. 3, pp. 231–239, 2009.

[PEE 14] PEETERS F., OEHLEN M., RONNER J. *et al.*, "Neurofeedback as a treatment for major depressive disorder–a pilot study", *PLoS ONE*, vol. 9, no. 3, 2014.

[PEN 89] PENISTON E.G., KULKOSKY P.J., "Alpha-theta brainwave training and beta-endorphin levels in alcoholics", *Alcoholism, Clinical and Experimental Research*, vol. 13, no. 2, pp. 271–279, 1989.

[PFU 00] PFURTSCHELLER G., GUGER C., MÜLLER G. *et al.*, "Brain oscillations control hand orthosis in a tetraplegic", *Neuroscience Letters*, vol. 292, no. 3, pp. 211–214, 2000.

[RAM 13] RAMOS-MURGUIALDAY A., BROETZ D., "Brain-machine interface in chronic stroke rehabilitation: a controlled study", *Annals of Neurology*, vol. 74, no. 1, pp. 100–108, 2013.

[ROC 90] ROCKSTROH B., ELBERT T., LUTZENBERGER W. *et al.*, "Biofeedback: Evaluation and Therapy in Children with Attentional Dysfunctions", *Brain and Behavior in Child Psychiatry*, pp. 345–357, 1990.

[ROC 93] ROCKSTROH B., ELBERT T., BIRBAUMER N. *et al.*, "Cortical self-regulation in patients with epilepsies", *Epilepsy Research*, vol. 14, no. 1, pp. 63–72, 1993.

[SCH 15] SCHARNOWSKI F., WEISKOPF N., "Cognitive enhancement through real-time fMRI neurofeedback", *Current Opinion in Behavioral Sciences*, vol. 4, pp. 122–127, 2015.

[SHA 13] SHARMA N., BARON J.-C., "Does motor imagery share neural networks with executed movement: a multivariate fMRI analysis", *Frontiers in Human Neuroscience*, vol. 7, 2013.

[SHE 11] SHERLIN L.H., ARNS M., LUBAR J. *et al.*, "Neurofeedback and basic learning theory: implications for research and practice", *Journal of Neurotherapy*, vol. 15, no. 4, pp. 292–304, 2011.

[SHI 11] SHINDO K., KAWASHIMA K., USHIBA J. *et al.*, "Effects of neurofeedback training with an electroencephalogram-based brain-computer interface for hand paralysis in patients with chronic stroke: a preliminary case series study", *Journal of Rehabilitation Medicine*, vol. 43, no. 10, pp. 951–957, 2011.

[SHO 79] SHOUSE M.N., LUBAR J.F., "Operant conditioning of EEG rhythms and ritalin in the treatment of hyperkinesis", *Biofeedback and Self-Regulation*, vol. 4, no. 4, pp. 299–312, 1979.

[STE 69] STERMAN M., LOPRESTI R., FAIRCHILD M., "Electroencephalographic and behavioral studies of monomethylhydrazine toxicity in the cat", 1969.

[STE 72] STERMAN M., FRIAR L., "Suppression of seizures in an epileptic following sensorimotor EEG feedback training", *Electroencephalography and Clinical Neurophysiology*, vol. 33, no. 1, pp. 89–95, 1972.

[STR 09] STREHL U., "Slow cortical potentials neurofeedback", *Journal of Neurotherapy*, vol. 13, no. 2, pp. 117–126, 2009.

[STR 14] STREHL U., BIRKLE S.M., WÖRZ S. *et al.*, "Sustained reduction of seizures in patients with intractable epilepsy after self-regulation training of slow cortical potentials – 10 years after", *Frontiers in Human Neuroscience*, vol. 8, no. 604, pp. 1–7, 2014.

[SUD 11] SUDRE G., PARKKONEN L., BOCK E. *et al.*, "rtMEG: a real-time software interface for magnetoencephalography", *Computational Intelligence and Neuroscience*, vol. 2011, 2011.

[SUL 13] SULZER J., HALLER S., SCHARNOWSKI F. *et al.*, "Real-time fMRI neurofeedback: progress and challenges", *NeuroImage*, vol. 76, pp. 386–399, 2013.

[TAN 09] TAN G., THORNBY J., HAMMOND D.C. *et al.*, "A meta analysis of EEG biofeedback in treatment of epilepsy", *Clinical EEG and Neuroscience*, vol. 40, no. 3, pp. 1–8, 2009.

[THA 98] THATCHER R., "Normative EEG Databases and EEG Biofeedback", *Journal of Neurotherapy*, vol. 2, no. 4, pp. 8–39, 1998.

[THA 13] THATCHER R., "Latest developments in live Z-score training: symptom check list, phase reset, and Loreta Z -score biofeedback", *Journal of Neurotherapy*, vol. 17, no. 1, pp. 69–87, 2013.

[WEI 03] WEISKOPF N., VEIT R., ERB M. *et al.*, "Physiological self-regulation of regional brain activity using real-time functional magnetic resonance imaging (fMRI): methodology and exemplary data", *NeuroImage*, vol. 19, no. 3, pp. 577–586, 2003.

[WOL 02] WOLPAW J., BIRBAUMER N., "Brain–computer interfaces for communication and control", *Clinical neurophysiology*, vol. 113, no. 6, pp. 767–791, 2002.

[WOL 11] WOLPAW J., WOLPAW E., "Brain–computer interfaces: principles and practice", 2011.

[WYC 14] WYCKOFF S., BIRBAUMER N., "Neurofeedback and brain–computer interfaces", *The Handbook of Behavioral Medicine*, pp. 275–312, 2014.

[WYR 68] WYRWICKA W., STERMAN M. B., "Instrumental conditioning of sensorimotor cortex EEG spindles in the waking cat", *Physiology & Behavior*, vol. 3, no. 5, pp. 703–707, 1968.

[YAN 08] YAN N., WANG J., LIU M., "Designing a brain–computer interface device for neurofeedback using virtual environments", *Journal of Medical and Biological Engineering*, vol. 28, no. 3, pp. 167–172, 2008.

[YOO 02] YOO S.-S., JOLESZ F.A., "Functional MRI for neurofeedback: feasibility study on a hand motor task", *Neuroreport*, vol. 13, no. 11, pp. 1377–1381, 2002.

[YOU 14a] YOUNG B.M., WILLIAMS J., PRABHAKARAN V., "BCI-FES: could a new rehabilitation device hold fresh promise for stroke patients?", *Expert review of medical devices*, vol. 11, no. 6, pp. 537–539, 2014.

[YOU 14b] YOUNG K. D., ZOTEV V., PHILLIPS R. *et al.*, "Real-time FMRI neurofeedback training of amygdala activity in patients with major depressive disorder", *PLoS ONE*, vol. 9, no. 2, 2014.

[ZOT 14] ZOTEV V., PHILLIPS R., YUAN H. *et al.*, "Self-regulation of human brain activity using simultaneous real-time fMRI and EEG neurofeedback", *NeuroImage*, vol. 85, pp. 985–995, 2014.

List of Authors

Fernando ARGELAGUET-SANZ
Inria Rennes-Bretagne Atlantique
Rennes
France

Christian BARILLOT
IRISA
CNRS
Rennes
France

Christian BÉNAR
INS, INSERM UMR S 1106
Aix-Marseille University
France

Laurent BOUGRAIN
LORIA
University of Lorraine
Vandoeuvre-lès-Nancy
France

François CABESTAING
CRIStAL
Lille 1 University
France

Géry CASIEZ
Lille 1 University
France

Maureen CLERC
Inria Sophia Antipolis-Méditerranée
Sophia Antipolis
France

Marco CONGEDO
GIPSA-Lab CNRS
Grenoble Institute of Technology
Grenoble-Alpes University
France

Emmanuel DAUCÉ
INS, INSERM UMR S 1106
Aix-Marseille University
France

José DEL R. MILLÀN
École Polytechnique Fédérale de
Geneva
Switzerland

Philippe DERAMBURE
CHRU de Lille
France

Andeol EVAIN
Inria Rennes-Bretagne Atlantique
Rennes
France

Rémi FLAMARY
Laboratoire Lagrange
Nice Sophia Antipolis University
France

Jérémy FREY
Inria Bordeaux-Sud-Ouest
Talence
France

Camille JEUNET
Inria Bordeaux-Sud-Ouest
Talence
France

Matthieu KANDEL
Centre Médical Germaine Revel
Saint Maurice sur Dargoire
France

Anatole LÉCUYER
Inria Rennes-Bretagne Atlantique
Rennes
France

Fabien LOTTE
Inria Bordeaux-Sud-Ouest
Talence
France

Jérémie MATTOUT
Centre de Recherches en
Neurosciences de Lyon
University of Lyon
France

Bernard N'KAOUA
University of Bordeaux
France

Lorraine PERRONNET
Inria Rennes-Bretagne Atlantique
Rennes
France

Christophe POUZAT
MAP5 and CNRS UMR 8145
University Paris-Descartes
France

Alain RAKOTOMAMONJY
Rouen University
Saint Etienne du Rouvray
France

Nicolas ROUSSEL
Inria Lille-Nord Europe
Villeneuve-d'Ascq
France

Raphaëlle ROY
ISAE Sup'Aéro
Aeronautics and Space Center
Toulouse
France

Michèle SEBAG
Inria Saclay-Ile de France
Gif-sur-Yvette
France

Maude TOLLET
CMPR Bel Air
La Membrolle-sur-Choisille
France

Franck VIDAL
Laboratoire de Neurosciences
Cognitives CNRS
Aix-Marseille University
Marseille
France

Index

J, K, L

M

N

O

P

R

S

T, U

V, W, X

Contents of Volume 2

Other titles from

iSTE

in

Cognitive Science and Knowledge Management

2016

FORT Karën
Collaborative Annotation for Reliable Natural Language Processing

GIANNI Robert
Responsibility and Freedom
(Responsible Research and Innovation Set – Volume 2)

LENOIR Virgil Cristian
Ethical Efficiency: Responsibility and Contingency
(Responsible Research and Innovation Set – Volume 1)

MATTA Nada, ATIFI Hassan, DUCELLIER Guillaume
Daily Knowledge Valuation in Organizations

NOUVEL Damien, EHRMANN Maud, ROSSET Sophie
Named Entities for Computational Linguistics

SILBERZTEIN Max
Formalizing Natural Languages: The NooJ Approach

2015

LAFOURCADE Mathieu, JOUBERT Alain, LE BRUN Nathalie
Games with a Purpose (GWAPs)

SAAD Inès, ROSENTHAL-SABROUX Camille, GARGOURI Faïez
Information Systems for Knowledge Management

2014

DELPECH Estelle Maryline
Comparable Corpora and Computer-assisted Translation

FARINAS DEL CERRO Luis, INOUE Katsumi
Logical Modeling of Biological Systems

MACHADO Carolina, DAVIM J. Paulo
Transfer and Management of Knowledge

TORRES-MORENO Juan-Manuel
Automatic Text Summarization

2013

TURENNE Nicolas
Knowledge Needs and Information Extraction: Towards an Artificial Consciousness

ZARATÉ Pascale
Tools for Collaborative Decision-Making

2011

DAVID Amos
Competitive Intelligence and Decision Problems

LÉVY Pierre
The Semantic Sphere: Computation, Cognition and Information Economy

LIGOZAT Gérard
Qualitative Spatial and Temporal Reasoning

PELACHAUD Catherine
Emotion-oriented Systems

QUONIAM Luc
Competitive Intelligence 2.0: Organization, Innovation and Territory

2010

ALBALATE Amparo, MINKER Wolfgang
Semi-Supervised and Unsupervised Machine Learning: Novel Strategies

BROSSAUD Claire, REBER Bernard
Digital Cognitive Technologies

2009

BOUYSSOU Denis, DUBOIS Didier, PIRLOT Marc, PRADE Henri
Decision-making Process

MARCHAL Alain
From Speech Physiology to Linguistic Phonetics

PRALET Cédric, SCHIEX Thomas, VERFAILLIE Gérard
Sequential Decision-Making Problems / Representation and Solution

SZÜCSAndras, TAIT Alan, VIDAL Martine, BERNATH Ulrich
Distance and E-learning in Transition

2008

MARIANI Joseph
Spoken Language Processing

Printed and bound by CPI Group (UK) Ltd, Croydon, CR0 4YY
16/09/2026
14963131-0001